THE MAHARISHI TECHNOLOGY OF THE UNIFIED FIELD:

The Neurophysiology of Enlightenment

by

Robert Keith Wallace, Ph.D.

Maharishi International University
MIU Neuroscience Press
Fairfield, Iowa

THE MAHARISHI TECHNOLOGY OF THE UNIFIED FIELD: THE NEUROPHYSIOLOGY OF ENLIGHTENMENT

9 8 7 6 5 4 3 2 1

Library of Congress Catalog Card Number: 86-061345

ISBN 0-9616944-0-8

Wallace, R. Keith

The Maharishi Technology of the Unified Field:
The Neurophysiology of Enlightenment

To Maharishi Mahesh Yogi

whose great wisdom, creative genius and continual guidance to help all men and women achieve enlightenment and to create world peace is an inspiration to all.

ACKNOWLEDGEMENTS

I am grateful to the ladies of Maharishi Vedic University for their valuable aid in editing. I would also like to thank Susan Shatkin for her conscientious and excellent effort in editing this book. In addition, I wish to express my thanks to John Kesterson whose superb organizing ability enabled all the elements of this book to be completed and printed. Thanks also to Claude and Manon Cazeault for typing and preparing the bibliography and charts, Shepley Hansen for the cover design, and Mike Davis for paste-up and layout.

I wish to acknowledge with gratitude the creative suggestions and contributions of Drs. John Hagelin, John Fagan, David Pasco, David Orme-Johnson, Michael Dillbeck and Laurence Domash.

My thanks to my mother and to my four wonderful children Eden, Ted, Gareth and Lila for their understanding and acceptance of my hours of research and writing.

I am especially appreciative of my wife's constant encouragement and support which made the long process of writing this book possible.

Contents

Introduction i

A Scientific Means to Develop Enlightenment.............. ii
The Unified Field of Natural Law iii
Neurophysiology of World Peace vi

Chapter 1: Physiology, Veda and Consciousness 1

Rediscovery of a Science of Consciousness—
Vedic Science 5
Higher States of Consciousness.................................. 9
The Transcendental Meditation Technique 13
The TM-Sidhi Program.. 15
The Scientific Discovery of Enlightenment 16

Chapter 2: The Discovery of the Unified Field of all the Laws of Nature 24

The Basis of the Physical Laws of Nature.................. 25
Breaking Apart the Atom... 26
Unified Field Theories 27
The Total Potential of Natural Law as
Located in Physiology 35
Self-referral DNA 37
Three-in-one Structure of DNA 40
DNA, the Material Expression of Samhita 43
Self-referral at Higher Levels of
Physiological Organization...................................... 45
An Integrated Science of Life 47

Chapter 3: The Physiological and Biochemical Correlates of the Least Excited State of Consciousness................................ 52

History of Research on Meditation Techniques 53

Early Studies on the Transcendental Meditation Technique —A Proposed Fourth Major State of Consciousness 56
Metabolism and Respiration 61
Lactate Generation 72
Redistribution of Blood Flow and Muscle Metabolism 75
Biochemical and Hormonal Changes 77
Autonomic Effects 84
EEG 85
EEG Coherence 88
Refinement of Methodologies to Study the Least Excited State of Consciousness 95

Chapter 4: The Development of Higher States of Consciousness 99

Homeostasis—Physiological Adaptability and Stability 101
Perceptual and Motor Performance Changes 108
EEG Trait Effects—Becoming Fully Awake 112
Biochemical Measurements 116
Psychological Measurements 120
Prison Rehabilitation Programs 122
Self-actualization 132

Chapter 5: Research on the TM-Sidhi Program 134

EEG Coherence 136
Sensory and Motor Responses 143
Biochemical Studies 148
Conclusion 153

Chapter 6: Perfect Health—Life in Accordance with Natural Law 156

Ayurveda: The Science of Life 157
The Relationship of Mind and Body in Health 158
Cardiovascular Risk Factors 159

Other Clinical Studies....171
Drug Abuse....175
Other Ayurvedic Approaches to Perfect Health....182
Psycho-physiological Body Types....183
Natural Medicine....190

**Chapter 7: Reversing the Aging Process—
Life in the Direction of Immortality....198**

Longevity Factors....199
Biological Aging....203
More Stable and Efficient Internal Physiology....210
Theories of Aging....212
The Potential to Oppose Aging in the Brain....216
Development of the Nervous System:
The Importance of Experience....218
Reorganization of Neural Pathways in
Higher States of Consciousness....219
Life in Accord with Natural Law....226

Chapter 8: Is Consciousness a Field?....229

Collective Consciousness....231
Raising the Coherence of Collective
Consciousness—Experimental Verification....233
The Super Radiance Effect....237
The Global Maharishi Effect—A Taste of Utopia....242
The Mechanism of the Maharishi Effect....254
The Physiology of World Peace....256

**Chapter 9: Neurophysiological Models of
Consciousness....261**
The Neurophysiological Basis of States
of Consciousness—Level 1....262
Neural Centers During Meditative States....266
Models of the Neurophysiology of
Transcendental Meditation....268
New Directions in Research on Pure
Consciousness—
Level 1 Theoretical Models....273

Theoretical Models—Level 2 280
Future Research—Level 2 Theoretical Models 286

Chapter 10: Education for Enlightenment 288

The Innate Physiological Ability to Experience Pure Consciousness 290
Revival of Knowledge 295
The Science of Creative Intelligence 301
Maharishi International University 304
Maharishi Vedic University 307
Education and Enlightenment—Maharishi's Integrated System of Education 311

Appendix I: Maharishi's World Plan for Perfect Health 316

Appendix II: Addresses and Contacts 323

Bibliography 326

Introduction

We are witnessing today a great transition in the history of mankind, a remarkable turning point in which the newest and oldest traditions of knowledge are at last converging and being unified into a new level of understanding and technology. While innumerable individuals have contributed to this transformation through their discoveries and insights, the primary recognition for this achievement of a rising new age of enlightenment, as history must show, goes to one man, Maharishi Mahesh Yogi. Over the last 30 years Maharishi, the founder of the Maharishi Technology of the Unified Field, has pioneered the development and understanding of an integrated science of life, which unifies the ideas of modern science, in particular the unified field theories of modern physics, with the complete wisdom of ancient Vedic Science. Further, Maharishi has introduced to over three million people of all cultures, religions, and educational backgrounds the practical aspect of this technology—the Transcendental Meditation and TM-Sidhi program. The result is the availability to all people everywhere of a technology for directly experiencing the unified field of all the laws of nature.

From my point of view as a physiologist, what Maharishi has accomplished is the single most important scientific discovery of our age, or for that matter any age. Perhaps there is no more appropriate time in history for such a discovery, a time when the laws of physics are on the verge of a long sought unification, and yet a time when the entire world lives in fear of total annihilation. For the first time in thousands of years, this ancient tradition of knowledge has been revived in completeness, a tradition that in fact includes a profound science of physiology. Until Maharishi began teaching it widely, this science of physiology was nearly extinct except in the practices of a secluded few individuals. It had for centuries been totally misunderstood by scholars and laymen alike in both the East and the West. The revival of this science has resulted in an unprecedented advance in our understanding of human consciousness and in the availability of a set of procedures for the development of an extraordinary state of neurophysiological functioning—a state traditionally referred to as enlightenment. As defined by Maharishi, the state of enlightenment is a state in which the awareness is established in the unified field of natural law, and in which activity and behavior are thus spontaneously in accordance with all the laws of nature. This state is achieved as a result of neurophysiological refinement and depends upon the perfect and harmonious functioning of every part of the body.

A SCIENTIFIC MEANS TO DEVELOP ENLIGHTENMENT

What is unique about this technology for the development of consciousness and physiology? First, it has been re-established in its purest and most effective form by Maharishi. Second, it is being expressed in a manner that makes it fully comprehensible and accessible in terms of the most recent theories and experimental procedures of physics, chemistry, mathematics, physiology and other disciplines of modern science. This meeting of ancient and modern science is removing the understanding of enlightenment from the realm of mysticism and uncertainty. It is showing enlightenment to be a scientific reality that is verifiable, universally available, and of immense practical value.

Enlightenment means, in a physiological sense, maximum orderliness and integration, perfect correlation between all aspects of physiological functioning—from the level of the DNA molecule, the total potential of natural law in living systems, to the highest expression of that potential in the functioning of the human nervous system. It represents the ultimate development of what we ordinarily consider the most valuable qualities of human life. It is something real and natural and develops systematically in a continuous and progressive manner on the basis of neurophysiological refinement, utilizing the existing mechanics of human physiology. Scientific research on the Transcendental Meditation and TM-Sidhi program has revealed several important features of the nature of enlightenment documenting marked improvements in all aspects of mind and body, including a reversal of the aging process. Taken together, this research begins to define physiologically the direction of enlightenment. Furthermore, it provides objective standards by which progress towards enlightenment may be measured. Through the Maharishi Technology of the Unified Field enlightenment can be achieved by anyone without adherence to a special life style or system of belief. The ability to gain enlightenment is innate in the physiology of every man and woman, and therefore every human being deserves to have the knowledge of how to utilize it. As Maharishi has said, "There is no reason today in our scientific age for anyone to remain unenlightened."

In this book we shall examine the Maharishi Technology of Unified Field, particularly from an experimental viewpoint, with reference to the many hundreds of scientific studies conducted at over 160 eminent universities and research institutes around the world. First, however, we will briefly introduce the concept of the unified field as it is understood from the perspective of the different disciplines of modern science and from the perspective of ancient Vedic Science.

THE UNIFIED FIELD OF NATURAL LAW

Within the last half of this century the various disciplines of modern science such as physics, mathematics, chemistry, and physiology have come to a common understanding that the basis of all physical activity is, in fact, an unmanifest or unexpressed field of knowledge. Modern science has demonstrated over and over that it is impossible to understand the full nature of a system by merely examining the superficial, excited levels of its activity. From the viewpoint of these more excited levels diversity, differentiation and change predominate, while from the perspective of the lesser excited and more unexpressed, underlying levels, integration, stability and unity predominate. The entire history of modern science is the history of the discovery of deeper levels of unification of the laws of nature, beginning with classical physics, developing into the more profound theories of quantum physics, and finally culminating in recent unified field theories.

The most profound aspect of the recent advances in modern physics is the description of an ultimate level of unification, the unified field of all the laws of nature, or super-field, which has the attributes of complete self-referral, self-sufficiency and infinite dynamism. These unique attributes of complete self-referral, self-sufficiency and infinite dynamism are precisely the same attributes that characterize the field of pure consciousness as described by Maharishi in his formulation of Vedic Science. Modern physics further describes a three-in-one structure of the unified field, in which the force and matter fields are united through the agency of supersymmetry. Likewise, in probing deeply into the ancient Vedic Literature, Maharishi describes a three-in-one structure of pure consciousness in which the knower and known are united by the process of knowing.

Thus modern physicists have concluded that the underlying nature of life is remarkably similar to the understanding expressed in ancient Vedic Science—it is a unified field, unmanifest, non-localized, infinitely dynamical, and has the qualities of self-sufficiency and quantum — sm. subdivision of energy

self-interaction or self-referral. The unified field contains the total potential of all the laws of nature in their most compact and integrated state. This achievement of modern physics—to have glimpsed a unified self-referral field of natural law at the basis of creation—is of immense importance, and is paralleled by similar discoveries in other fields of modern science such as mathematics, chemistry and physiology.

In mathematics we find that the axioms of set theory describe the null or empty set as an unmanifest, self-sufficient and self-referral field of intelligence. From this unmanifest null set, the full range of all mathematical theories, from the finite to the infinite, can be sequentially generated. Similarly, in modern chemistry, the basis of all the diverse chemical reactions and processes is located in the unmanifest quantum mechanical nature of the transition state. The transition state is a self-referral field of all possibilities, which ultimately gives rise to all possible behaviors of any chemical system. In modern physiology the source of all physiological structures and functions is located in the metabolically silent DNA molecule. All physiological processes are ultimately referred back to the knowledge within DNA. DNA acts in effect as the self-referral, self-sufficient source of all biological knowledge in living systems.

The objective approach of modern science has made an enormous advance in describing the nature of reality as an unmanifest field of natural law. However, as Vedic Science reveals, complete knowledge of this ultimate field can only be gained on the level of consciousness through direct subjective experience. This is because of the completely self-referral nature of this field. It is a field of pure consciousness—pure knowledge—and therefore can be known only by itself, by consciousness aware of its own unified nature. Modern science by its very emphasis on the object has excluded from its investigation the subject—the knower—and the process of knowing. Only when these three components—knower, known and process of knowing—are experienced in their completely unified state can complete or pure knowledge be realized.

To achieve this unique experience of pure consciousness being aware of its own nature, one must develop a refined level of neurophysiological functioning. The Maharishi Technology of the Unified Field includes a very specific process designed to culture the nervous system and the entire physiology so it can support lesser and lesser excited states of consciousness until finally it is able to support and spontaneously maintain the least excited state of consciousness—pure consciousness. The ability to maintain this state of pure consciousness gives direct experience of the unified field, the total potential of natural law, and enables individual awareness, and consequently all individual thinking and action, to be in accord with all the laws of nature at all times.

NEUROPHYSIOLOGY OF WORLD PEACE

The implications of this new technology are far reaching, extending beyond the individual to include the whole of society. According to Maharishi, at the basis of individual consciousness is a field of collective consciousness that underlies the coherent behavior of society. By aligning individual consciousness with the unified field it is possible not only to bring individual life in accord with natural law but also to positively influence the overall quality of life in all areas of society.

A peaceful individual is the unit of world peace. The unit of world peace, in turn, is structured in the specific pattern of neurophysiological functioning that is generated by the Maharishi Technology of the Unified Field. We thus have in our possession a technology to structure the neurophysiology of enlightenment for each individual and also to structure the neurophysiological basis of world peace. Given the current world situation, it is very timely that Maharishi has offered the practical means to create world peace, to create a unified field based civilization.

It has always been the characteristic of great scientific discoveries to produce unforeseen technological breakthroughs of immense benefit to human life. It is the very great fortune of this age to have been given this gift of enlightenment by Maharishi. It is also the great responsibility of this age to utilize this precious knowledge and technology revived by Maharishi in order to eliminate, as quickly as possibly, suffering and violation of natural law, and to usher in an age of enlightenment.

Chapter 1

PHYSIOLOGY, VEDA AND CONSCIOUSNESS

For thousands of years man has tried to understand the nature of consciousness and its relationship to the laws of nature. Until less than 300 years ago virtually all of the methods for gaining knowledge were subjective in nature. At the core of almost every cultural tradition of mankind can be found specific subjective techniques which attempt to enliven an inner intuitive understanding of natural law. This is particularly true in the Vedic tradition of India, where refined technologies of meditation were developed to probe into one's own nature, into the Self, the state of pure consciousness. Even at the beginning of our Western tradition, in Socrates' famous dictum "know thyself," the emphasis was to understand man's essence first, for it was felt that man was but a representation, or microcosm, of the entire universe, and that by knowing the nature of the inner self the essence of natural law might be realized. As a result, life could be lived more in accord with natural law.

The modern approach to understanding the nature of consciousness is the objective methodology of science, which in the last 300 years has become the dominant and effective tool of inquiry into nature. The neurosciences are those disciplines that have been concerned specifically with utilizing this objective methodology to study the nervous system and brain, those physical structures that support the integrity of consciousness. Throughout the history of neurophysiology the primary methodology has been reductionistic in nature: an attempt to localize the subjective nature of consciousness in specific structures or processes within the brain. In so doing a number of questions have been raised, including: What is the physical basis of memory? What are the anatomical structures involved in emotional behavior? How can we define mental health in terms of the balance of specific chemicals in the brain?

While we have gained a great deal of information concerning the basic electrophysiological and biochemical mechanisms of such simple neurophysiological systems as the nerve axon or neuromuscular junction, we still know very little about the neurophysiological basis of more integrated neural systems, let alone the human brain itself. Many students of neuroscience enter this field with a desire to gain a more complete picture of the neurophysiological basis of consciousness. However, after a few years in graduate school, most typically find themselves investigating the electrical or chemical properties of neural membranes in squids or snails, an area of basic importance, but far removed from their original intention to better understand consciousness.

The Nobel laureate David Hubel, in an introduction to a collection of articles entirely devoted to current research on the brain, poses the question, "Can the brain understand the brain? Can it understand the mind? Is it a giant computer or some other kind of giant machine, or something more?" These questions are fundamental not just to research on the brain but to the whole direction of science.

Most brain researchers today, I think, feel that the brain is only beginning to understand the brain. While this beginning is itself one of the most exciting and significant areas of human knowledge, a new direction is clearly needed. Another Nobel laureate, Francis Crick, in an article entitled "Thinking About the Brain," points out that "...the brain is clearly so complex that the chance of being able to predict its behavior solely from a study of its parts is too remote to consider. We sense there is something difficult to explain, but it seems almost impossible to state clearly and exactly what the difficulty is. This suggests that our entire way of thinking about such problems may be incorrect." In his article Crick emphasizes the importance of new approaches, for, as he says, "There is no scientific study more vital to man than the study of his own brain. Our entire view of the universe depends on it."

The human brain is indeed the most precious gift of nature. It is the link between the abstract subjective world of consciousness and the concrete objective realm of physical matter. The famous neuroanatomist Ramon y Cajal said, "As long as the brain is a mystery, the universe, the reflection of the brain, will also be a mystery." How can we understand the brain, and more importantly how can we understand the nature of consciousness?

We can gain an insight into what steps are in fact necessary for a new direction or breakthrough in neuroscience by looking into one of the great breakthroughs in modern science: the formulation of quantum physics, and with it the realization that there exist in fact two very different types of physical reality—the reality of classical physics and the reality of quantum physics. In the classical realm we assume that particles and events are concrete, localized and predictable. Furthermore, they can be observed and measured in space and time, and our observation is independent of the measurement itself; that is, it does not change the thing we are observing.

In quantum physics the situation is entirely different. We cannot picture the quantum realm, for it is based on the behavior of systems that cannot be clearly observed. The particles and events are abstract, nonlocalized, and in fact are described in terms of probabilities. Further, the quantum world is not independent of the observer, but rather very dependent upon the act of observation, as demonstrated by the famous Heisenberg uncertainty principle. Thus there are two separate realities, the manifest classical world and the unmanifest quantum world.

This realization by modern physicists of two separate realities should also, I believe, be applied to modern neuroscience. The first reality or level is the nervous system; the electrical and biochemical activities of the nervous system are localized and measurable. The second reality is the mind or consciousness, which is abstract and nonlocalized in its nature and difficult to measure directly.

The great success of quantum theory lies in its ability to describe the very fine levels of matter far more accurately than classical physics. It does not discard the observations of classical physics, but rather reveals that the laws of classical physics are correct only on a macroscopic scale where matter and energy exist in excited states. The more fundamental states can be properly described only by the more encompassing and powerful language of quantum field theory. Thus, while the laws of classical physics are in agreement with those of quantum physics in the more excited states of matter they are unable to describe the more fundamental, lesser excited states.

I believe the same situation exists in the field of neuroscience. The present concepts and approaches are classical in their nature and are limited in their ability. They reflect the activities of only the most excited states and are unable to probe into the lesser excited, more fundamental states of the nervous system and of consciousness. A new approach in neuroscience, akin to that taken by quantum theory, is needed, one which deals with the phenomenon of consciousness more directly. This approach, therefore, could more accurately describe the lesser excited states of the nervous system without discarding all of the great achievements of classical neuroscience.

In order to create such a new field in neuroscience that deals with consciousness more directly we must, however, have available a research tool or technology that is sophisticated enough to study the nervous system in its purest and least excited states. The Maharishi Technology of the Unified Field is just such a research tool. While the origin of this technology, the Vedic tradition of India, is both unfamiliar and unexpected to most scientists, extensive research over the last decade has revealed its immense importance to both science and society.

REDISCOVERY OF A SCIENCE OF CONSCIOUSNESS—VEDIC SCIENCE

Over the last 30 years Maharishi Mahesh Yogi has developed a science and technology of consciousness that provide an entirely new approach and insight into the nature of consciousness. Consciousness, according to Maharishi, is not merely an individual human subjective experience or stream of awareness, but is the most fundamental field of nature, the unified field of natural law. As we have briefly seen and will elaborate in greater detail in the next chapter, Maharishi's description of the field of pure consciousness is virtually identical to the description of the unified field or super-field of modern physics. This new science of consciousness, the Maharishi Technology of the Unified Field, has theoretical, experimental and applied values in all areas of life. The theoretical and experimental components are derived from three principle sources of knowledge and research: 1) the ancient Vedic tradition of India, 2) the Transcendental Meditation and TM-Sidhi program and 3) the principles and discoveries of modern science. Let us briefly examine the first of these sources as an introduction to this new field of research.

The Vedic tradition of India is considered by many scholars to be the oldest living tradition of knowledge preserved by man. Unfortunately, the interpretation of this knowledge has been confined, until now, to a very superficial level. This is primarily because Vedic knowledge can be properly understood only from the perspective of higher states of consciousness, and to develop these higher states, a precise set of procedures is needed. Without an effective methodology to experience higher states of consciousness, the essential meaning of the Vedas cannot, according to Maharishi, be realized.

It is Maharishi's genius to have revived this profound knowledge. With extraordinarily profound insight, he has reinterpreted the Vedic literature in terms of the experience of higher states of consciousness and demonstrated that far from being stories or hymns, they are a remarkably precise description of the dynamical principles underlying the laws of nature.

The word Veda refers to the state of pure knowledge or pure consciousness in which the knower, process of knowing and known are completely unified. Vedic Science, as formulated by Maharishi, provides a very complete and detailed elaboration of this unified state of pure knowledge. Further, it systematically describes a first principle of the inherent dynamism of natural law. This is the principle of self-referral or self-interaction of knower, process of knowing, and known, through which this three-in-one structure gives rise to the vast diversity of natural law.

As a result of this first principle of the self-referral dynamism in nature, consciousness in its self-interacting state becomes aware of itself, leading to the emergence of three distinct components within the one fundamentally indivisible field of pure consciousness. Proceeding in one direction, the one (consciousness) assumes the structure of three (knower, process of knowing, known); and in the other direction, three in turn converge to become one. This fundamental dynamism establishes an inherent pulsation in the unmanifest structure of pure consciousness, which underlies all the expressions of natural law in creation. In the words of the Rig Veda, "*Richo Akshare parame vyoman, Yasmin deva adhi vishve nisheduh.*" "The verses of the Veda exist in the collapse of fullness (the "kshara" of A or inherent pulsation of the universe) in the transcendental field in which reside all the devas, impulses of creative intelligence, the laws of nature responsible for the whole universe."

The four books known as Rig Veda, Sama Veda, Yajur Veda and Atharva Veda serve to record the words of Vedic literature. However, they themselves are not the Veda. Veda is the field of pure potentiality, pure intelligence, pure knowledge, that indestructible, immortal level of reality which contains all the impulses of natural law structuring the rest of creation. How, then, can one know the Veda? Maharishi refers to a verse from the Rig Veda to answer this question:

> *Yastanna veda kimricha karishyati; Ya ittadvidusta ime samasate.*" "He whose awareness is not open to this field, what can the verses or impulses of pure knowledge accomplish for him? He whose awareness is open to it, he is established in evenness—wholeness of life."

> The impulses of natural law can be cognized only in the state of pure intelligence, pure knowledge. They cannot be known except in a very superficial way in ordinary, excited, waking state consciousness. Their real value can be realized only in the least excited state of consciousness, the state of pure consciousness—the simplest form of human awareness, the Self. In this state consciousness appreciates its own fine structure moving within itself.

> The practical value of this state of pure knowledge, consciousness aware of itself, is that grounded in this state the individual begins to act in accord with natural law. Established in this state of wholeness, every impulse of action arises from an impulse of natural law, and thus all activity is most effective and life-supporting. The whole process is very spontaneous, as the Rig Veda explains.
>
> *Yojagaratam richah kamayante.*
>
> The richas (impulses of natural law) seek out him who is awake.

To be fully "awake," according to Maharishi, means to be established in pure consciousness, pure knowledge. This is a prerequisite for gaining knowledge of the Veda. Once one is fully awake no effort is needed; in this state of pure awareness the nature of the underlying threads of pure awareness begin to reveal themselves to themselves. Therefore the Veda is structured in consciousness, and is accessible through the Maharishi Technology of the Unified Field, which refines neurophysiological functioning such that human awareness is open to the direct experience of consciousness in its pure self-interacting state.

The nature of Vedic cognition is such that the knower, process of knowing and known remain unified within the state of pure consciousness. The richas are experienced not as separate from oneself, but as the modes of one's own intelligence—consciousness reverberating within itself. Thus, the Veda is the unified state of the knower, the process of knowing and the known—referred to in the Vedic literature as Samhita—and its three essential components are known as rishi, devata and chhandas respectively.

These essential components of experience—rishi, devata and chhandas—are separately elaborated in the extensive body of Vedic literature, including the Brahmanas, Vedangas, Upangas, Itihasas, Puranas, Smritis and Upavedas. The various branches of the Vedic literature are seen to emerge from the Samhita, from which they derive their vitality and authenticity. In fact, all the disciplines of modern science can also be seen to have their origin in the Samhita which represents the internal structure and dynamics of the unified field itself.

HIGHER STATES OF CONSCIOUSNESS

According to the Vedic tradition of knowledge, what we consider normal waking consciousness is in fact a very limited experience of consciousness that is confined to the more excited levels of the mind. Systematically quieting the internal physiology and at the same time enlivening mental awareness allows the experience of a least excited or ground state of consciousness. This ground state of consciousness is referred to as the state of transcendental or pure consciousness, in which there are no thoughts, no sensory experiences, and no distinction between subject and object—only pure awareness, the experience of consciousness itself.

The Vedic tradition not only describes a ground state of consciousness, which is distinctly different from the waking, dreaming or sleep states of consciousness, but, as we have mentioned, further specifies the existence of a set of "higher" or more optimal states of consciousness. These higher states of consciousness are characterized by the simultaneous coexistence of the ground state —pure consciousness—along with the waking, dreaming, or sleep states. Furthermore, they give rise to a more perfect or optimal state of neurophysiological functioning.

Why have these descriptions of meditative procedures and higher states of consciousness contained in the Vedic texts, which have such obvious practical advantages, been so ignored and mistrusted in Western and even Eastern civilization? To answer this question, we must keep in mind one critically important fact: the practice of these techniques develops and perfects the physiology of the participants. The knowledge of how to attain higher states of consciousness, and eventually enlightenment, includes an ancient science or technology of physiology, which contains many procedures designed to purify and refine the body and nervous system. Greater physiological refinement in turn supports deeper subjective experiences. The state of enlightenment thus depends upon, and is defined in terms of, the perfect and harmonious coordination of mind and body, a unique physiological state.

Precisely because enlightenment depends upon a unique physiological state, it becomes difficult to achieve when the technology for gaining it easily is lost. Maharishi explains that this ancient science, which included meditation procedures, was universally known in the time of ancient Vedic civilization in India, but was gradually lost except to a handful of individuals, "due to the long lapse of time." The technology became distorted because of a lack of proper understanding of how to fully utilize the procedures to refine and perfect the nervous system. Once these systematic methods were lost, there arose in their place an enormous variety of less effective and more difficult techniques. Meditation procedures no longer contained their once universal character, but were replaced by austere practices of renunciation and detachment. Over the course of time the high regard once given to higher states of consciousness and the state of enlightenment was replaced by mistrust and disbelief.

Maharishi, in rediscovering and developing the precise and effective methodology to refine the physiology in a simple, natural, and effortless way, has enabled the experiences and principles contained in the Vedic tradition to be correctly understood and independently experienced by hundreds of thousands of individuals throughout the world.

How did Maharishi revive this knowledge? He continually gives credit for this revival to his teacher, His Divinity Swami Brahmananda Saraswati, Jagad Guru Bhagawan Shankaracharya (affectionately referred to by Maharishi as "Guru Dev") who was regarded throughout India as a unique living example of the highest level of inner perfection as expressed in the Vedas.

In India, the traditional means of gaining knowledge is through oral instruction passed from teacher to disciple under the strictest supervision. The tradition from which Guru Dev comes can be traced back many thousands of years to the great Indian saint and philosopher Shankara. The line of Shankara in turn has its origin in the Vedic tradition. Thus the tradition of knowledge Maharishi has revived extends in an unbroken continuum from earliest Vedic culture to modern times. It had for centuries been principally a monastic and reclusive tradition. Even among those who were accepted into it only a small number were recognized as having achieved complete enlightenment; they often lived in solitude inaccessible to all but a few disciples.

In 1941, following a long period in which the principle seat of the Vedic tradition had been unoccupied, Swami Brahmananda Saraswati, in answer to repeated requests by India's most revered philosophers and scholars, left his reclusive life of solitude and assumed the position of the Shankaracharya of Jyotir Math of Northern India. During this time Maharishi became his closest disciple and undertook training in the tradition that for countless generations has preserved and protected in its purity the essence of Vedic knowledge, which includes the procedures for the refinement of physiological functioning and the attainment of enlightenment.

Maharishi spent some thirteen years under the guidance of Guru Dev, and when his teacher passed away in 1953, he went into seclusion in Northern India. After several years Maharishi left seclusion temporarily to make a pilgrimage to South India. The journey, however, became considerably longer than he had expected. While he was in South India, he was asked to lecture on the teachings of his master. After this first lecture the demand became so great for the knowledge Maharishi had to offer that he spent the next thirty years, and continues even now, lecturing and teaching around the world. Maharishi took an unprecedented step: he made available on a large scale—to men and women from all walks of life and in all parts of the world—a teaching that had for many years been known only to a handful of monks.

According to Maharishi's interpretation of the most renowned texts of Vedic literature (such as the Bhagavad Gita) there were, originally, two types of meditation techniques, one for monks and one for householders. The procedures for monks had lost their effectiveness over time, but had been maintained, at least to some extent, within the walls of the monasteries. The householder procedures, however, were completely lost. It was assumed that one had to first lead a monk's life before one could learn to meditate. Maharishi revived both types of procedures in their full effectiveness. Further, he taught that anyone could begin the householder procedures without first having to change his or her lifestyle. These householder procedures, which constitute the experiential component of the technology of the unified field, are commonly known as the Transcendental Meditation or TM program and the more advanced TM-Sidhi program.

THE TRANSCENDENTAL MEDITATION TECHNIQUE

The mental image evoked by the word "meditation" is that of an ascetic monk sitting cross-legged in a remote mountain cave, with little protection or clothing and virtually no possessions. To most of the western world, such a lifestyle obviously lacks appeal. Not only would the specific practice of an austere technique of meditation be extremely demanding and time-consuming, but the prerequisite social and ethical codes would be both incongruous and incompatible with the social and materialistic pattern of life in modern civilization.

The TM technique is a unique, simple and effective mental procedure. Most people have begun the technique for such practical reasons as the desire for better physical and mental health, increased energy, decreased anxiety and tension, more fulfilling family relations, improved clarity of thought, and greater success and fulfillment. The technique involves no mood, belief, or specialized lifestyle; rather it involves a real and measurable process of physiological refinement. It utilizes the inherent capacity of the nervous system to refine its own functioning and to unfold its full potential. In a spontaneous and natural way during the practice, the attention is drawn to quieter, more orderly states of mental activity until all mental activity is transcended, and the observer is left with no thoughts or sensations, only the experience of pure awareness lively in itself.

While the experience of transcending can be verbally described, (and many beautiful literary descriptions of it exist), it is an experience that depends upon a very specific and delicate state of the nervous system. It cannot properly be learned through books, but must be learned through personal instruction by a trained teacher. This helps prevent forcing, straining or expectation which, according to Maharishi, hold the mind on a gross level and disallow the experience of more delicate and subtle levels of the thinking process.

Most systems of meditation teach that the tendency of the mind is to wander, and that in order to experience quieter levels of consciousness, the mind must be controlled. Maharishi explains that the tendency of the mind is not to wander aimlessly, but to move naturally in the direction of experiences that bring greater happiness and enjoyment. He states, and this is the common experience reported by meditators, that increasingly quieter and more refined levels of thinking are progressively more enjoyable. Therefore, one needs to know only how to begin the procedure for allowing finer levels of thought—the more delicate stages of the process of the development of a thought—to come to conscious awareness and spontaneously the attention is drawn inward. While this procedure is completely natural to the mind, it is also very delicate. Without proper instruction and checking by a qualified TM teacher, misunderstandings in meditation can unnecessarily arise, and the individual may not gain the full benefits of the practice.

Personal instruction is necessary for another reason. Although, in principle, any thought could be chosen as a vehicle for transcending, the Vedic tradition, from which the Transcendental Meditation technique is derived, has systematically found that certain particular thoughts or sounds are most conducive to producing maximum beneficial effect. The scientific verification for the appropriateness of these sounds, or "mantras," is based primarily on thousands of years of the experience of the teachers of this tradition. That the sound value of a thought is chosen rather than its meaning or its value as a visual image is interesting and important in itself.

According to Maharishi, the sound value is chosen because we are most commonly aware of thought as a sound. We "hear" our thoughts in our mind. At subtler levels, the thought may actually acquire polysensory characteristics, that is, it is simultaneously heard, visualized, felt, and so forth. From the point of view of neurophysiology, this is important for it suggests the involvement of more integrative areas of the brain such as the frontal cortex, which receive input from all the major senses. In chapter 9 we will develop these ideas in greater depth as we explore various models of the neurophysiological basis of transcending.

THE TM-SIDHI PROGRAM

In 1976, approximately 15 years after the introduction of the TM technique, Maharishi began teaching a set of advanced procedures known as the TM-Sidhi program. These procedures were derived from that part of Vedic literature known as the Yoga Sutras of Patanjali. Patanjali's sutras have been available in translation for a number of years, yet their interpretation and application are still misunderstood. Maharishi recognized that their effectiveness was entirely dependent upon the ability to transcend and maintain the state of pure consciousness, which occurs naturally a result of the regular practice of the TM technique. It is a misinterpretation of Patanjali's original instructions to attempt these procedures without the ability to transcend.

The TM technique cultures the nervous system to maintain the state of pure consciousness, a state referred to by Patanjali as samadhi. The TM-Sidhi program involves applying Patanjali's sutras or formulas in a very specific way as taught by Maharishi. The procedure provides a unique means of stirring the field of pure consciousness, of transforming the quiet nature of pure consciousness into a dynamic self-sufficient field. Using the sutras in this way develops extraordinary psychophysiological functioning and mind-body coordination. More importantly, it stabilizes the experience of pure consciousness, thus accelerating growth to higher states of consciousness.

Already many thousands of individuals, a number of whom are professionals in the fields of medicine, law and education, have begun this new area of research in the expansion of human potentialities. Perhaps the most interesting of the TM-Sidhi procedures is the Patanjali sutra for "flying." Although the full phenomenon of flying has not yet been achieved, the technique has proven itself to be the most powerful of the TM-Sidhi practices in terms of laboratory physiological measurements and personal experience.

The rediscovery of the effectiveness of these procedures and the development of an educational program by which they can be taught to thousands of individuals throughout the world illustrate Maharishi's extraordinary abilities as both a world teacher and a profound research scholar.

THE SCIENTIFIC DISCOVERY OF ENLIGHTENMENT

Maharishi's ability to present these Vedic techniques of self development in a scientific manner has encouraged researchers at leading universities around the world to conduct hundreds of physiological, psychological, sociological, and ecological studies on the effects of the Transcendental Meditation and TM-Sidhi program.

The results of these studies can best be understood in terms of four fundamental questions that underlie the relationship between consciousness and the laws of nature. These four questions are:

1. Is there a least excited state of consciousness and a corresponding least excited state of the nervous system?

2. Are there higher, more optimal states of consciousness, and corresponding higher levels of organization and more optimal states of neurophysiological functioning?

3. Is there a field of collective consciousness that underlies the orderliness and coherence of social behavior?

4. How does this underlying field of pure consciousness influence the physical laws of nature?

The following chapters of this book are devoted to answering each of these questions in detail. Let us consider them briefly now.

The first question can be considered from a theoretical and an experimental viewpoint. From a theoretical perspective we can examine the characteristics of the least excited state of consciousness as developed by Maharishi from the Vedic literature to see if there are equivalent descriptions of the theories of modern science, in particular modern physics. From such an analysis, as we have briefly discussed and shall explore more fully in the next chapter, we can identify three basic qualities of the state of pure consciousness: complete self-sufficiency, self-referral, and infinite dynamism. We might refer to these qualities as the dynamical properties of consciousness. These properties of consciousness are indeed quite similar to the dynamical qualities of matter and energy as formulated by modern physics, particularly as seen in the description of the super-field in unified field theories.

From an experimental perspective the existence of a least excited state of consciousness can be examined by attempting to characterize the physiological and biochemical correlates of this state. The initial research on the TM technique and a great deal of subsequent research has concerned itself with this identification. These studies suggest that during the TM technique a state of restful inner alertness is produced. This state is characterized by a number of specific physiological and biochemical changes such as a marked reduction in oxygen consumption and carbon dioxide elimination, an increase in skin resistance, a decrease in arterial blood lactate and plasma cortisol, and a marked increase in EEG alpha and theta coherence in the frontal and central areas of the brain.

These first studies I shall refer to as phase I studies, because, while they describe the general characteristics of a wakeful hypometabolic state, they are limited in one very important aspect—they do not make the critical distinction of whether the measurements made corresponded specifically to the subjective experience of the least excited or pure state of consciousness or in fact corresponded to a mixture of this state with more excited states of wakefulness and normal relaxation. Fortunately, a new set of studies has recently been conducted which I shall refer to as phase II studies. These studies reveal that during the TM technique there may be a mixture of states, i.e., wakefulness, pure consciousness, and drowsiness. Care must be taken to distinguish which substate is being studied and which measurements are used, some being more appropriately sensitive to discriminate between substates. When these criteria are met it is possible to more clearly identify the profound physiological correlates of the state of pure consciousness.

Let us turn now to a brief elaboration of the second of the four questions previously mentioned: Are there higher, more optimal states of consciousness, and correspondingly, are there higher levels of organization and more optimal states of neurophysiological functioning? Another way of asking this question is: Does the experience of the state of pure consciousness during the TM technique positively influence states of consciousness and physiological functioning outside of meditation?

Indeed, a number of studies have shown extremely beneficial changes outside of the TM technique, such as faster recovery from stressful stimuli; more stable and lower resting levels of basic physiological functions; improvements in a number of cardiovascular risk factors (including reduction of blood pressure and high cholesterol levels); improvement in other disorders such as asthma and insomnia, as well as in mental health; faster reaction time; and better performance on perceptual motor tasks, learning and memory tests, and a variety of psychological inventories.

Several recent studies have demonstrated that the regular practice of the more advanced TM-Sidhi program is accompanied by longitudinal changes in neurophysiological function which are correlated with an increase in intelligence, creativity, concept learning, and moral reasoning. Researchers have suggested that the higher levels of EEG coherence in TM-Sidhi participants are a result of a more integrated style of brain functioning which allows the individual to interact in a more flexible and harmonious manner with his social and physical environment.

Further, if we examine physiological and biochemical studies conducted on the TM and TM-Sidhi program from the point of view of the aging process, we find that in virtually every case this program produces changes that are in a direction opposite to those which characterize aging. In one study, the biological age of a number of subjects practicing the TM and TM-Sidhi program was measured. It was found that the long-term meditators had a significantly younger biological age (12 to 15 years younger than their chronological age) than non-meditating controls and norms for the general population. In addition there was also a significant correlation between younger biological age and the length of time practicing the TM and TM-Sidhi program. These and other studies taken together thus clearly define more optimal levels of physiological functioning which are associated with the development of higher states of consciousness.

The third question we have posed is even more far reaching than the first two: Is there a field of collective consciousness that underlies the orderliness and coherence of social behavior?

If such a field of consciousness exists, it should be possible to test it by measuring its field properties. For example, certain physical systems (for example, lasers) exhibit properties such that if there is a coherent subpopulation of a small number of elements then the system undergoes a phase transition and begins to display macroscopic coherence. Applying this principle to society we might predict that if consciousness is indeed a field, a small coherent subpopulation of individuals could generate a more widespread coherent influence on the whole of society. This coherent influence could then be measured by changes in specific social indices. This approach has been undertaken in a number of studies.

Maharishi predicted a number of years ago that when one percent of the population of a society practiced the TM program a measurable improvement, such as a decrease in crime rate, would occur in the quality of life of that society. This effect has been observed in a number of different studies conducted in populations of various sizes. For example, in one study by Dillbeck and coworkers (1981), crime rate trend in 48 different cities was analyzed over a 12 year period. The 24 experimental cities, defined by having one percent of the population practicing the TM program, showed a significant decrease in crime rate trend as compared to 24 control cities randomly selected from matched cities with similar economic, educational and other demographic characteristics. This decrease in crime rate trend in the "one percent" cities has been shown to be independent of such factors as police coverage, unemployment, prior crime trend, difference in age composition, and ethnic background. This field effect has been appropriately called the Maharishi Effect.

An even more powerful effect has been noted with group practice of the TM-Sidhi program. This effect, known as the Super Radiance Effect, requires only the square root of one percent of a population to produce measurable effects such as reduction of violence and increased economic prosperity. Twenty studies have documented the effectiveness of the Super Radiance Effect in improving the quality of life in numerous cities around the world. The results of these studies cannot be accounted for unless one considers consciousness to be a field which is capable of transmitting effects over long distances. The discovery of these effects is of fundamental importance since it demonstrates both the existence of an underlying field of pure consciousness and, more importantly, a profound technology which, by coherently enlivening this field, can improve all areas of life.

The fourth and final question follows naturally from the previous three. How does this underlying field of pure consciousness influence the physical laws of nature? The relationship between consciousness and physical matter is a recurring theme in the history of science. Most recently, in the development of modern quantum field theory, a number of physicists have commented on the importance of a more complete understanding of human consciousness and the role it plays in quantum measurement. For example, the Nobel laureate Eugene Wigner states, "Our inability to describe consciousness adequately, to give a satisfactory picture of it, is the greatest obstacle to our acquiring a rounded picture of the world." Bernard D'Espagnat expresses a similar perspective: "The doctrine that the world is made of objects whose existence is independent of human consciousness turns out to be in conflict with quantum mechanics and facts established by experiments."

Maharishi, in his reinterpretation of Vedic knowledge,

describes very clearly the relationship between consciousness and physical laws. He states that the least excited or pure state of consciousness underlies both mental phenomena as well as physical phenomena and represents a unified level of natural law.

The theoretical and practical applications of this understanding of consciousness for neuroscience are far reaching. In one particular model postulated by Domash (1976), consciousness is seen as a global non-localized phenomenon. Drawing a parallel to the third law of thermodynamics, the TM technique is viewed as a methodology for the conscious exploration of a very low "mental temperature." Domash suggests by analogy to physical systems that when the "mental temperature" or internal noise level reaches its lowest level of excitation a phase transition to a distinct and more highly ordered state occurs within the nervous system.

The TM-Sidhi procedures expand this model further. As Maharishi explains, these procedures provide a means by which individual consciousness can come into contact with the unified field of natural law and directly stimulate the impulses of the laws of nature within that field. These impulses of natural law can be seen as the dynamic principles that govern the flow of consciousness. These principles are realized not in the ordinary sense, in terms of understanding specific mathematical relationships, but instead by bringing individual awareness directly in alignment with natural law on the level of consciousness itself.

By refining neurophysiological functioning and thereby establishing the least excited state of consciousness permanently in our awareness, we gain direct access to the unified field of all the laws of nature, the total potential of natural law. When we act from this level of awareness, all our activity becomes spontaneously in accord with natural law, restoring balance to all physiological functions and promoting ideal health and behavior.

The Maharishi Technology of the Unified Field thus offers modern science not only the opportunity to understand the relationship between the nervous system and the underlying field of pure consciousness, but more importantly provides a practical methodology for improving the quality of life on the level of both the individual and society.

Chapter 2

THE DISCOVERY OF THE UNIFIED FIELD OF ALL THE LAWS OF NATURE

You place matter before life and you decide that matter has existed for all eternity. How do you know that the incessant progress of Science will not compel scientists to consider that life has existed during eternity, and not matter? You pass from matter to life because your intelligence of today cannot conceive things otherwise. How do you know that in ten thousand years one will not consider it more likely that matter has emerged from life?

— Louis Pasteur

Intelligence is the basic value of all creation, and of all processes of progress and evolution in creation. It is the fundamental of all existences. This field of pure intelligence, which is one, non-dual, by virtue of its perpetual, eternal, immortal existence, starts to regenerate itself through its own nature, by virtue of its own existence. Intelligence becomes creative intelligence, and creates from its own nature.

— Maharishi

In the last ten years modern physics has made enormous achievements in understanding the fundamental basis of the different forces in nature, resulting in the development today of a unified field theory which promises to provide a complete understanding of natural law. In this chapter we will briefly trace the progress of modern quantum theory and consider in detail the unified field of all the laws of nature. Further we will examine the development of a unified understanding of natural law from the perspective of other scientific disciplines, particularly modern physiology. Throughout this discussion our main consideration will be to theoretically locate the total potential of natural law, the least excited state of both matter and consciousness. This will enable us to formulate the theoretical basis of an integrated science of life.

THE BASIS OF THE PHYSICAL LAWS OF NATURE

What are the physical laws that govern matter? What in fact is the nature of physical matter itself? At the dawn of Western civilization these were fundamental questions posed by the Greek philosophers. In the fifth century, the Greek philosophers Lisipicus and Democritus expressed the concept of the atom as the basic building block of universe—the smallest indivisible and indestructible substance of matter. Unlike earlier theories, which inevitably took on some form of animism, Democritus' view clearly divided spirit from matter. Matter was seen as something that had no life, which was as if dead, moved only by some invisible forces or laws of nature yet to be fully discovered and described.

Some 2,000 years later, as a result of the genius of such men as Galileo and Newton, these early philosophical ideas of matter were revived and formulated into a mechanistic world of indestructible particles of matter, particles that God had constructed out of material so hard and solid that they would never wear or break. These particles were acted upon by external forces according to certain laws of nature, which could be described by mathematical equations. Newton's view of the world persisted from the second half of the seventeenth century to the end of the nineteenth century. This mechanistic billiard ball model of the universe completely dominated scientific ideas and created a deterministic view of nature, which permeated all philosophical, educational, and social thought. The reality, or underlying substance of life, was perceived as being solid and material.

BREAKING APART THE ATOM

In the last century, several major discoveries in modern physics have completely shattered the classical Newtonian model of the universe. First, matter was found to be neither solid nor static. It was discovered that the atom is in fact mostly empty space in which invisible electrons whirl at velocities of approximately 600 miles per second around a tiny nucleus. The high velocity of the electrons in the atom give it the appearance of solidity, just as a rapidly rotating propeller might appear as a solid disc. Within the nucleus itself are particles that also move about at extremely high velocities—up to 400,000 miles per hour. The nucleus, being very dense, contains almost all the atom's mass. In fact, if the human body were compressed to nuclear density, it would take up less space than a pinhead.

A second major discovery which came out of the very revolutionary theory of quantum mechanics is that matter has a dual aspect—it appears sometimes as particles and sometimes as waves. One of the consequences of the wave aspect of matter is that it is impossible to have a well-defined measurement of the momentum or velocity of the particle and at the same time to have a well-defined, precise measurement of its position. The mathematical form of this relationship between the uncertainty of the position and momentum of the particle is well known among scientists as the Heisenberg uncertainty principle. We may choose to measure with precision either position or momentum, but we may never know the measurement of both at the same time. Thus, the more deterministic models of classical physics were forced to give way to a dynamic, ever-changing description of the universe.

A third important discovery, expressed in Einstein's theory of special relativity as the famous equation $E=mc^2$, is that matter is not indestructible, but is rather a form of energy that may be transformed into other forms of energy or matter.

We can no longer think of an atom as a solid particle; it is mostly empty space. Matter has a dual aspect—particle and wave. Further, matter is no longer seen as something dead but as a condensation of vibrating energy. In reality, solid matter is only a condensation or a concentrated manifestation of an underlying unmanifest field of energy and intelligence that permeates and upholds everything in creation.

UNIFIED FIELD THEORIES

According to modern physics, the essence of matter is its underlying field. In the words of Albert Einstein:

> We may therefore regard matter as being constituted by the regions of space in which the field is extremely intense...
>
> There is no place in this new kind of physics both for field and matter, for the field is the only reality. The field is seen as a continuum. It is present everywhere in space, and when we want to think of a particle, we have to think of it as being a blemish or discontinuity in the field structure.

For a number of years it was understood that there were four fundamental forces: electromagnetism, weak, strong, and gravity. These forces are described in quantum field theory as force fields. Over the past decades a progressively more unified understanding of the fundamental force fields has occurred, culminating in the recent discoveries of unified field theories. In 1967 the first stage came with the unification of the weak and electromagnetic forces by Weinberg, Salam and Glashow. According to their theory, for which they received the Nobel prize in 1979, at a fundamental scale of nature's functioning (at a distance of 10^{-16} cm which is 100 times smaller than the nuclear dimensions), the weak and electromagnetic forces become indistinguishable. Several years after they received the Nobel prize, experimental confirmation for their theory came with the discovery of two new particles, the W and Z particles.

This unified electro-weak theory established the importance of the so called "unified non-Abelian gauge theories" and the principle of spontaneously broken symmetry, which locates deep symmetries of nature at fundamental space-time scales. An important property of non-Abelian gauge theories is self-interaction or self-referral. The electromagnetic field by itself does not possess this property: protons do not interact with protons. It is only at deeper levels of unification in quantum field theory, where the electromagnetic field is united with the weak force field, that this property of self-referral, that is, the field's ability to interact with itself, appears.

Following the development of the electro-weak theory it was discovered that the principle of broken symmetry could be extended to include the strong force. This led to the development of grand unified theories, in which the weak, electromagnetic and strong force were unified, at a distance scale of 10^{-29} cm. Finally, in 1974, the profound unifying concept of supersymmetry was introduced, making it possible to unite matter fields (fermi fields) with force fields (bose fields) at the level of the Planck scale, 10^{-33} cm. This ultimate level of unification has led in recent years to the introduction of supergravity and supersymmetric unified field theories, such as extended supergravity theory and heterotic superstring theory.

From the perspective of these unified field theories, the property of self-interaction or self-referral reaches its ultimate degree of self-sufficient expression at the Planck scale, the level of unification where the extreme self-interaction of quantum gravity leads to a phase transition in the structure of space-time itself. At this level there is neither space nor time, only an infinitely dynamic sea, or "superfoam," as it is called. This level of unification is believed to be the dimension of the physical universe at its very inception some 20 billion years ago.

Once the universe began to cool, moments after the big bang, it expanded, and diversity arose on the manifest level of creation. Through the process of spontaneous sequential dynamical symmetry breaking the supersymmetry of the four forces was broken and gravity was separated from the other three. At progressively larger distance scales this process continued so that the remaining three forces, in turn, separated. This process of sequential dynamical spontaneous symmetry breaking refers to the way in which the symmetry breaking is initiated—from within the supersymmetry itself. By a dynamic process of self-referral, the supersymmetry transforms itself into its expressed parts—the fundamental laws of nature.

Thus all the laws of nature can be seen to be contained in their unmanifest, unexpressed state in the unified field. At this level, as we have discussed, we find the properties of complete self-sufficiency, self-referral and infinite dynamism.

This picture is precisely equivalent, as we saw in

Chapter 1, to the descriptions of natural law as revealed by Maharishi in his revival of Vedic Science. The state of pure consciousness is defined by Maharishi as being completely self-sufficient, self-referral, and infinitely dynamic. Maharishi elucidates, in his Apaurusheya Bhashya commentary on the Rig Veda, the mechanics of the unfolding of natural law in Vedic Science, the way in which the structure of natural law is generated from within consciousness. This process is the same as that described by the expression "sequential dynamical spontaneous symmetry breaking". Step by step, consciousness reveals within itself the complete structure of the laws of nature—as a self-expressed commentary of the full potential of natural law, which resides in its unified state in pure consciousness.

Maharishi's commentary identifies the property of self-interaction or self-referral as a first principle of the dynamics of nature's functioning. As we have described, according to Maharishi's formulation of Vedic Science, the Veda is defined as the unified state of knower, known and process of knowing—the state of pure knowledge or pure consciousness. This three-in-one structure of the Samhita, or unified state of the Veda, is remarkably similar to the structure of the unified field in which the bose fields (force fields) and fermi field (matter fields) are unified through the agency of supersymmetry.

Dr. John Hagelin, a leading physicist in unified field theories, has pointed out that the bose field corresponds to the "observer" aspect of the unified field (the knower), the fermi fields to the material or "observed" aspect (the known), and the agency of supersymmetry as the "process of observation" between the two (the process of knowing). The self-referral dynamics of pure consciousness lead to the emergence of this fundamental three-in-one structure in the same manner that the process of spontaneous symmetry breaking leads to the emergence of the three-in-one structure of the unified field as described by physics.

Through repetition of this process, a succession of unmanifest impulses of natural law is generated in a manner similar to that described by physicists to have occured at the beginning of the universe. These impulses of natural law interacting within themselves ultimately result in the manifestation and formation of curved space-time (Hagelin, 1984).

In Maharishi's analysis of the Rig Veda, each word and verse is seen to systematically elaborate on the mechanics of the unfolding of creative intelligence—the mechanics of creativity within consciousness itself. Maharishi explains that the entire mechanics of creativity for the whole universe is contained in the first word of the Rig Veda, Agnim, which describes the dynamics of spontaneous symmetry breaking. In discussing the Sanskrit etymology of each of the first four letters of Agnim, the nature, origin, range, and development of creative intelligence is clarified.

The first syllable, A, represents the totality of knowledge of the entire Veda, fullness. Fullness moving within itself locates in its own nature unmanifest emptiness, expressed by the letter G. As Maharishi explains, fullness would not be full if it did not contain even unmanifest emptiness within itself. Upon locating emptiness within itself, fullness, as if shocked, recoils from the emptiness. This is expressed by the letter N, signifying negation. In this process of negating and recoiling, the unmanifest value of fullness is stimulated to manifest and creation begins.

The development of creation, expressed in the letter I, represents the progression of unmanifest emptiness coming back towards the fullness of A. Finally, M represents the infinite dynamics of consciousness moving within itself, the dynamics which structure all creation. These dynamics are further elaborated in the rest of the

PHYSICS

This chart illustrates how the unified field of natural law is the source of all the fundamental force and matter fields in nature, and how these fields in turn provide the basis for all areas of physics and technology and their applications throughout society. The Maharishi Technology of the Unified Field unites the knowledge of the unified field discovered by the objective approach of modern physics with the direct experience of the unified field provided by the subjective approach of ancient Vedic Science. It enlivens the unified field in the awareness of the individual, bringing thought and action spontaneously in accordance with natural law so the individual enjoys the full support of nature in every aspect of life.

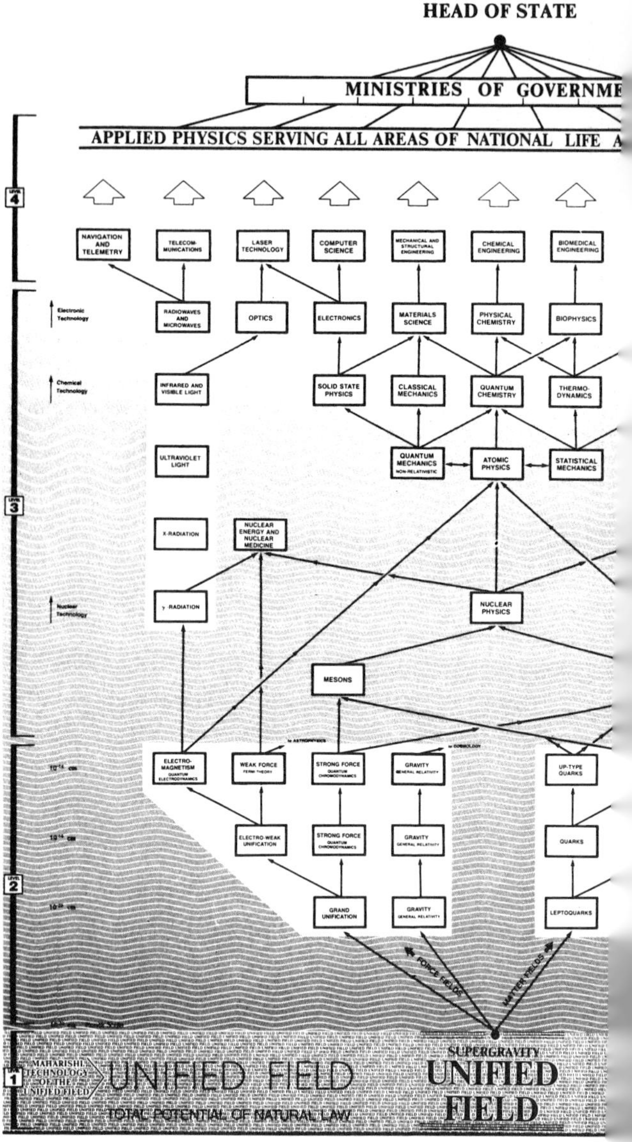

TRANSCENDENTAL MEDITATION

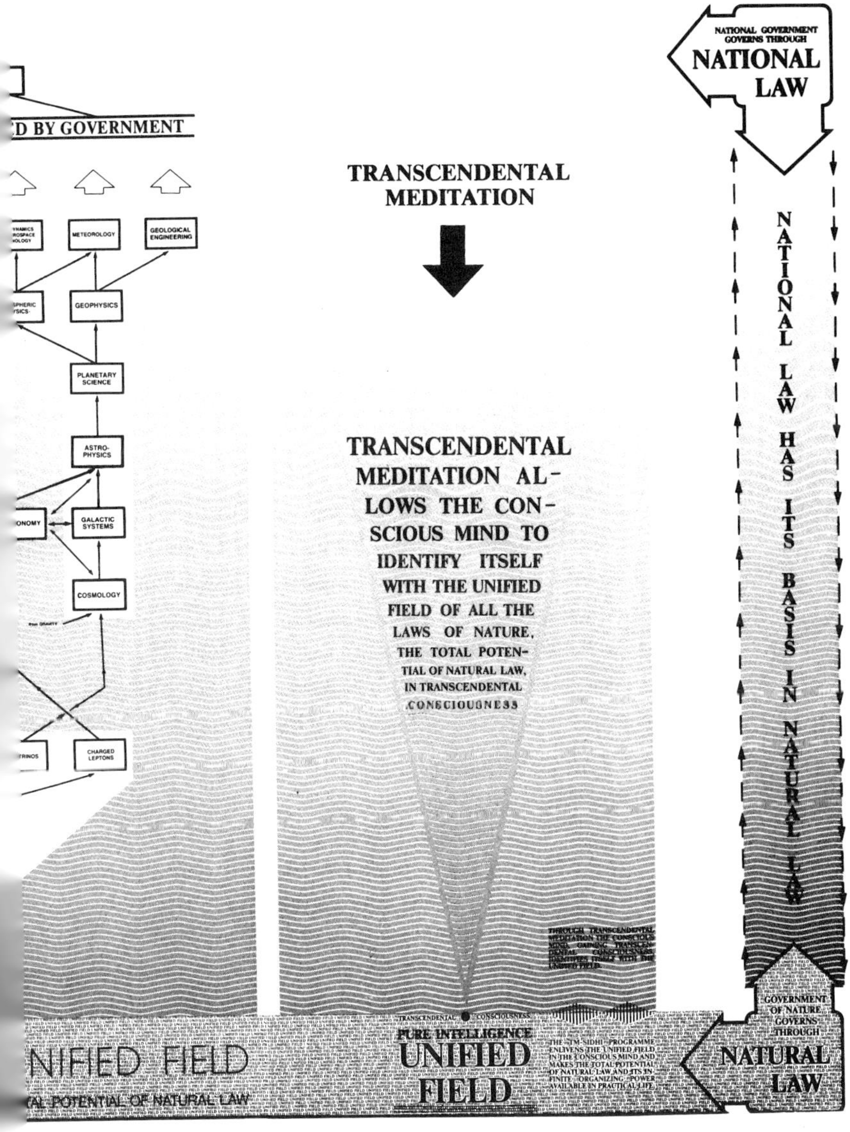

first richa of the Rig Veda.[1]

In this stepwise manner the entire Rig Veda is expressed. The rishi, devata, and chhandas values of the Rig Veda as a whole are then elaborated respectively in the Sama, Yajur and Atharva Vedas. These values are further sequentially elaborated in the entire Vedic literature. The Veda stands as its own self-expressed commentary.

Maharishi explains this process:

> Imagine a mass of deep silence, if it could become aware of itself, what it would find would be silence. According to the laws of quantum mechanics, in becoming aware of itself, the observer (silence) observing itself (silence) should have done something to itself. What will have happened is that the moment the awareness realizes, "I am silence," both an observer and an object of observation are created, and in this process space, time, motion are generated. The unmanifest provides the impetus for creation just by

---

1. The 24 gaps between the syllables of the first richa are, in turn, elaborated as the 24 padas in the remaining eight richas of the first sukta. Each pada contains eight syllables, so there are 8 x 24 = 192 syllables in the remaining eight lines of the first sukta. The 192 gaps between the 192 syllables of the first sukta find expression in the 192 suktas of the first mandala of the Rig Veda, 191 suktas plus one avyakta (unmanifest) sukta. The syllables of the second through ninth richas of the first sukta find elaborated expression in the second through ninth mandalas, while the gaps between the 192 suktas of the first mandala give rise to the 192 suktas of the tenth mandala.

> becoming aware of itself. This is the start of the entire time-space geometry—the beginning of creation. From consciousness creation begins.

Vedic Science as rediscovered by Maharishi makes a profound contribution to modern physics in that it not only enriches the description of the unified field to include the principle of self-awareness or self-consciousness, but further provides the technology whereby this level of creation can be directly experienced by each individual.

THE TOTAL POTENTIAL OF NATURAL LAW AS LOCATED IN PHYSIOLOGY

In the introduction we briefly examined the latest findings in mathematics, chemistry, and physiology and saw how each one also locates the total potential of natural law in an unmanifest unified field which stands as the basis of the discipline. Here we will elaborate this principle more fully in the field of physiology. We will do so by considering the origin of physiology from the viewpoint of modern physics, in particular unified field theories. By tracing the evolution of life from the unified field, with reference to a branch of physics known as non-equilibrium thermodynamics, it is possible to view evolution in terms of the unique property of self-referral.

The evolution of the universe, according to unified field theories, is a consequence of this first dynamical property of nature, the self-interactive nature of the unified field. As we have seen, proceeding from deeper levels of unification to the more expressed level of the four separate fundamental forces of nature,this property of self-interaction or self-referral becomes less apparent. At larger distance scales, it is as if it disappears. It re-emerges again only when we reach the macroscopic level of integrated physical, chemical and biological systems.

The field of non-equilibrium thermodynamics, as developed by Prigogine, gives perhaps the most descriptive analysis of this property of self-referral or self-interaction (Prigogine and Glansdorff, 1971; Prigogine et al., 1972).

According to non-equilibrium thermodynamics the evolution of any system, whether physical, chemical or biological, involves the dynamical self-interaction of different modes or fluctuations of the system. This process of self-interaction depends upon both the nature of the flow of energy through the system and the specific boundary conditions of the system. Under certain conditions, the fluctuations of a system will interact in such a manner as to eventually stabilize and form a new, more stable structure which is better adapted to the environmental conditions present. For example, when water is drained from a bathtub its flow causes turbulence or fluctuations. The fluctuations are a result of the self-interaction of hydrostatic forces. If the conditions are right, eventually a whirlpool is formed at the drain. The whirlpool is a delicate but stable structure which enables the water to flow out more quickly.

According to this theory, the same principles involved in the formation of the whirlpool are also responsible for the formation of life. Energy must be constantly flowing through the system and there must be a creative self-interaction between the different forces in the system. The more complex and integrated the system, the more apparent is this property of self-interaction or self-referral.

The underlying level of intelligence which enables apparently random self-interactions to form complex systems, ultimately is encoded in the unmanifest laws of nature contained in the unified field, the laws of nature which regulate and maintain all of creation. In physiological systems, this underlying level of intelligence which guides all the complex biological interactions in living systems is encoded in a particular physical structure, the DNA molecule. Let us more closely examine DNA as the self-referral unified basis of physiological systems to see if we can draw parallels to the mechanics of creativity as seen in modern science and Vedic Science.

SELF-REFERRAL DNA

Virtually every cell in most living organisms contains DNA, and within the DNA is encoded the totality of knowledge of the whole organism. The information in the DNA controls both the holistic development and maintenance of physiological functions. The DNA is, in effect, the unified basis of all the laws of nature for all physiological activities, in the same manner as the unified field is the underlying basis for the entire universe. Further, just as the unified field is characterized by the property of ultimate self-interaction or self-referral, DNA is also characterized by this property.

On the simplest level we can see the quality of self-referral in the basic structure of DNA, in which there is complementary base pairing between the double helical strands of nucleotides. The complementary base pairing is a form of self-referral in which one specific nucleotide always refers to its complementary pair. Complementary base pairing, among other attributes, provides an exquisite means of preserving the integrity of the information in DNA both by creating a more stable structure and also by facilitating the accuracy and efficiency of self-repair mechanisms.

Another example of self-referral occurs when one strand of DNA folds back on itself and assumes a specific spatial configuration known as a hairpin. This fold-back structure plays a key role in a number of regulatory functions (e.g., maintaining the stability of certain eukaryotic chromosomes, allowing double precise excision of transportable elements and signaling the termination and even

This chart illustrates how the Maharishi Technology of the Unified Field contributes to modern physiology by providing a new integrated approach in which the whole range of physiology can be appreciated from its source in the self-interaction of the self-referral unified field of all the laws of nature. Practice of the Maharishi Technology of the Unified Field ensures that the homeostatic, self-regulatory processes functioning at every level of physiology fully reflect the self-referral value of the unified field, establishing all aspects of life in accordance with the total potential of natural law—perfect health for the individual and society.

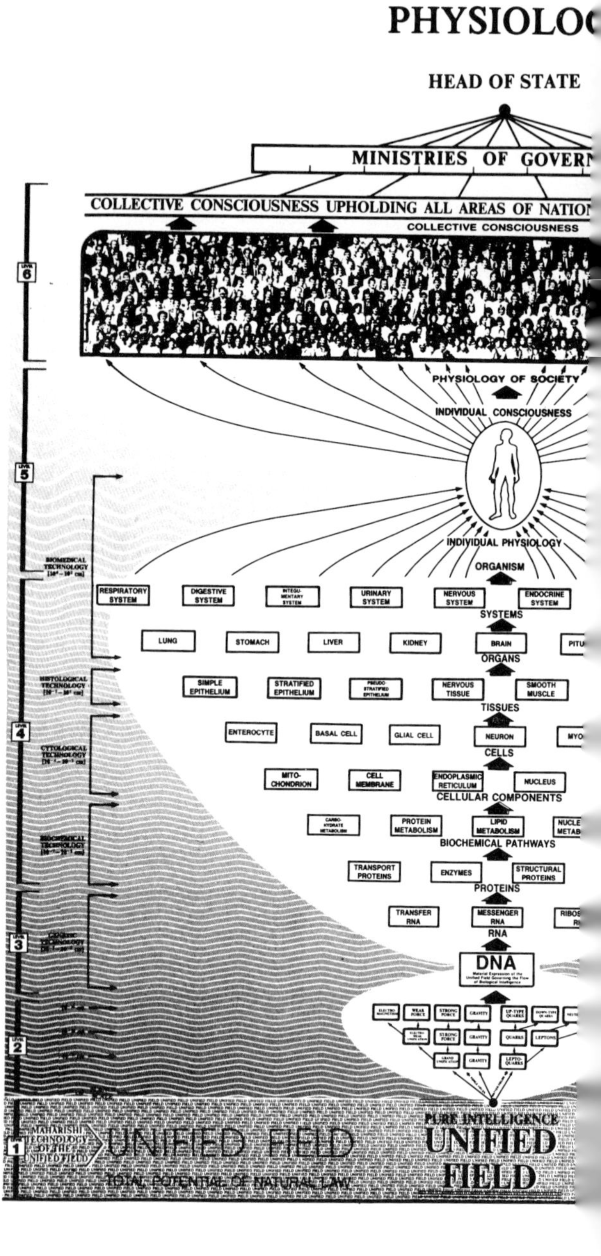

TRANSCENDENTAL MEDITATION

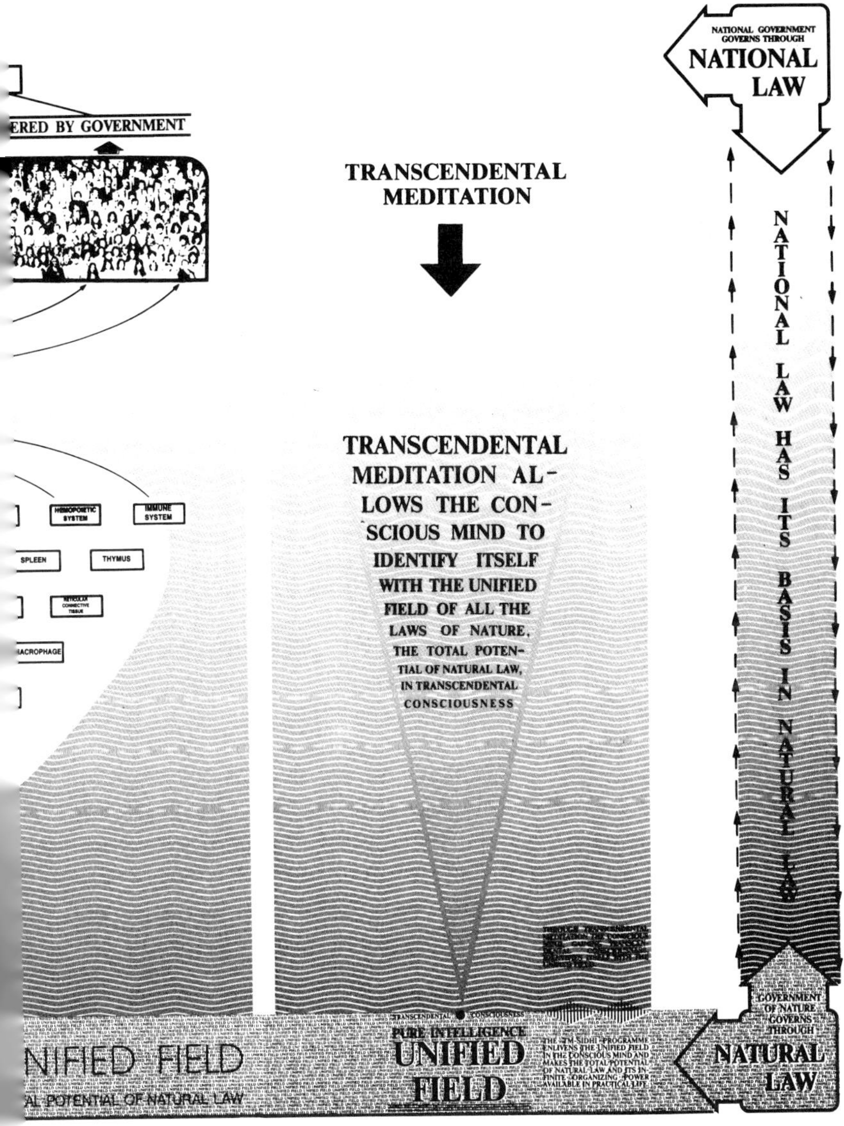

initiation of transcription). Such interactions of DNA within itself inevitably result in a breaking of the original symmetry of DNA's double helical structure, a process which is necessary for the expression of the information within it.

Perhaps the most obvious example of self-referral in the functioning of DNA is seen in the process of gene regulation. In gene regulation, regulatory molecules such as steroids, hormones, metabolites, and embryonic organizers, which are ultimately created by the knowledge in DNA, refer impulses of physiological information back to the DNA either directly or in cooperation with regulatory proteins. This may lead to activation or inactivation of particular genes, and in turn, specific physiological responses. The huge variety and precision of such responses, as seen in the development of the human body from a single cell, would be incomprehensible without knowledge of the self-referral mechanisms of gene regulation which govern the balanced unfoldment of genetic information from DNA.

THREE-IN-ONE STRUCTURE OF DNA

In considering the self-referral mechanics by which DNA expresses itself we can identify a three-in-one structure, similar to that described in modern physics and Vedic Science. In terms of DNA we can consider the unified or Samhita value as being the genome, the totality of knowledge contained in DNA. The rishi (knower), devata (process of knowing), and chhandas (known) can be seen to correspond to genetic information, the process of complementary base-pairing and the sequence of nucleotides in the DNA molecule, respectively. Just as rishi, devata, and chhandas are simultaneously locatable at every point in the Samhita of the Veda—the self-referral three-in-one structure of pure knowledge—genetic information, complementary base-pairing and nucleotide sequence are all present at every point in the genome.

On the level of the genome, the knower—rishi—is the genetic information. Molecular biology speaks of genetic information as the sum total of information specifying every aspect of biochemical, cellular, and organismic structure and function, including the information specifying the regulation and control of the expression of those structures and functions. It is clearly understood in molecular biology that this information is not the DNA molecule itself, but is stored in the medium of the DNA molecule in the same way that information is stored in a book in the media of paper and ink.

Complementary base-pairing is the process of knowing—devata— on the level of the DNA molecule. Just as the information in a book becomes lively in the awareness of the knower through the visual process, genetic information is made lively within the DNA molecule through intermolecular interactions, principally complementary base-pairing. On the level of the genome, complementary base-pairing creates a tight self-referral loop between the two strands of the DNA molecule that enlivens the genetic information, but leaves it in its pure knowledge form. In the process of gene expression, complementary base-pairing occurs between the template strand of the DNA and RNA precursors, resulting in the transfer of genetic information from the genomic DNA to messenger RNA (mRNA) molecules. Subsequently, complementary base-pairing occurs between the mRNA and transfer RNA (tRNA) molecules, to express the genetic information carried in the mRNA as a protein molecule.

In the genome, the nucleotide sequence is the known—chhandas. Genetic information is carried in the physical structure of the nucleotide sequence of the DNA molecule just as information is carried in a book in the sequence of letters on the page. The nucleotide sequence is also not identical to the physical structure of the DNA molecule, but is carried in the medium of the DNA molecule, just as language is not identical to the paper and ink of the book, but carries information in the medium of the book.

The study of DNA in the light of Vedic Science gives us the broadest possible perspective in which the value of intelligence, or knowledge, is seen to be the fundamental basis of physiology. DNA is seen as the knower, process of knowing, and known of all physiological processes.

DNA is the knower because it is the ultimate recipient or experiencer of all the impulses of knowledge of the environment. Because DNA's most immediate environment, the physiology, is nothing other than an expression of the DNA, the pure knowledge contained within itself, DNA is actually experiencing or knowing itself. Knowledge of increasingly expanded values of the environment is experienced by DNA through the physiology. However, the relationship of the physiology with its environment is also stored within the knowledge in DNA. Therefore, all knowing by DNA is of itself.

DNA is also the process of knowing. The knowledge within DNA is responsible for all aspects of the dynamics of experiencing its environment, and this experience is essentially self-referral. These quotes from Erwin Schrödinger (1945) and Maharishi Mahesh Yogi (1985) bear upon this point:

> ...The chromosome structures are at the same time instrumental in bringing about the development they foreshadow. They are law-code and executive power or, to use another simile, they are architect's plan and builder's craft in one....
>
> — Schrödinger

> DNA is the knowledge and organising power in one—rishi, devata and chhandas in one.
>
> — Maharishi

Further, DNA is the known, because the knowledge within it is responsible for giving rise to its most immediate environment, the physiology. DNA is also the known in that it contains knowledge of the environment beyond the physiology. This is so because the relationship between the physiology and increasingly expanded values of its environment determines the characteristics of the physiology, and therefore the knowledge expressed from the DNA. This is clearly observed in the evolution of the enormous variety of physiological structures in different species, which enables each in its own specific way to better adapt to a particular niche in the environment.

DNA, THE MATERIAL EXPRESSION OF SAMHITA

DNA can be seen as the material expression of the Samhita—the unified state of rishi, devata, and chhandas. It is present in virtually all the trillions of cells of the human body. In each case, the totality of knowledge in DNA is contained within each cell. In terms of the activity of any given individual cell only a small portion of the total information in DNA needs to be expressed.

The DNA has in effect both an individualized and collective role in regulating the physiology. The expression of specific information controls the highly specialized activity of certain individual cells. The coordinated expression of knowledge in an aggregate of cells controls the coordinated activity of tissues and organs. Finally, the totality of information in DNA acts as an underlying field of knowledge which upholds the overall collective coherence and integrity of the whole body.

In the remarkable process of physiological development, the complete knowledge or blueprint of the body, initially contained in the DNA of one single cell, is gradually and sequentially unfolded to form, through the processes of differentiation and morphogenesis, all the hierarchical layers of physiological structure and function.

During differentiation, the dividing cells become committed or specialized to perform a specific task (e.g., liver cells, heart cells, etc.) and as a result only express a small fraction of the complete knowledge which is contained in the DNA. During morphogenesis, there is an orderly migration and replication of cells to form the innumerable structures of the body. The entire orchestration of this highly complex process, which may involve the intricate interconnectedness of literally billions of cells, occurs as a result of the precise and sequential expression of knowledge from DNA.

This presents a marvelous picture of the Samhita of the Veda and its sequential emergence into speech and physiology. The Samhita of the Veda is present at every point in creation; it is the underlying field of pure knowledge from which all diversity emerges.

The unfoldment of the whole range of Vedic literature is a process of elaboration of the fundamental qualities of the Samhita—rishi, devata, and chhandas. During this developmental process, the complex body of Vedic literature sequentially unfolds as syllables or sounds, the impulses of natural law reverberating within the field of pure intelligence, the unified field of all the laws of nature. The continual self-referral dynamics of these impulses of intelligence give rise to the classical manifest world of matter and ultimately to the DNA. Thus, the name or sound value of the Veda gives rise to the form or matter value of creation. The sound value of the Veda is perfectly reflected in the mechanics by which the name or linear sequence of nucleotides in the DNA gives rise to the three dimensional form or matter of proteins. It is knowledge which ultimately structures matter.

SELF-REFERRAL AT HIGHER LEVELS OF PHYSIOLOGICAL ORGANIZATION

As we examine the more expressed levels of physiology arising from the knowledge in DNA we can also see the self-referral mechanics of DNA present at higher levels of physiological organization in the form of higher level homeostatic feedback systems. Homeostasis refers to the ability of living systems to maintain internal orderliness and stability in the presence of changes in their external environment through self-regulating feedback systems. The principle of homeostasis is perhaps the single most important concept in physiology. All the various biochemical and cellular processes, as well as the function and structure of tissues, organs, and organ systems are governed by self-referral, homeostatic mechanisms.

Every homeostatic system has a type of three-in-one structure: (1) an integrative center which acts as a kind of knower, witnessing and comparing incoming information to some predetermined set point; (2) a mechanism to transfer information, corresponding to the process of knowing; and (3) an effector mechanism through which the final output is expressed, corresponding to the known.

Homeostatic feedback mechanisms are essential to the regulation of vital parameters such as temperature, blood pressure, acid-base balance, water and electrolyte balance, oxygen and carbon dioxide levels, and glucose levels. Particularly elegant examples are seen in the endocrine and nervous systems, which regulate the overall functioning of the other systems of the body. The degree to which the homeostatic mechanisms maintain greater physiological balance and self-organization is reflected in the ability of the system to remain free from disorder and aging.

By creating internal stability, homeostatic mechanisms form the basis for the reliable and orderly flow of information from the fundamental level of DNA to the complex functioning of the nervous system. Established in the knowledge of DNA—its own internal source of stability and order—physiological activity is spontaneously balanced and in accord with the total potential of natural law.

This locates in physiology the universal principle of action in nature as described in Vedic Science:

> *Yogastah Kuru Karmani* (Bhagavad-Gita, II, 48). Established in the unified field of all the laws of nature—the total potential of natural law—perform action.

The establishment of individual awareness in its own self-referral state—the state of pure awareness, the ground state of all the laws of nature—brings every aspect of life in accordance with natural law. The result of this alliance with natural law through the Maharishi Technology of the Unified Field is to maintain all aspects of life in an evolutionary direction.

Viewing physiology from the broadest perspective, all individual aspects can be seen as the knowledge within DNA interacting and moving within itself. Thus DNA can be seen to display the properties of self-sufficiency, self-referral, and infinite dynamism, which are the attributes of consciousness. Such a unified perspective in physiology becomes a concrete and practical reality only when the full potential of DNA is unfolded. This occurs when the functioning of the human nervous system, with its billions of interconnecting neurons—the highest expression of the knowledge within DNA—is brought to a state of perfect coherence and complete self-referral through the TM and TM-Sidhi program.

The development of perfect physiological coherence is achieved in the state referred to as unity consciousness, in which all aspects of natural law are directly experienced in terms of the unified field of natural law, cosmic intelligence, which is the Self of the individual.

> In this state, consciousness is its own physiology.
>
> — Maharishi

In the attainment of unity consciousness, the quest of science to understand and utilize the total potential of natural law is fulfilled. This is the real fulfillment of modern physiology. When modern science is combined with the knowledge and technology of Vedic Science, an integrated science emerges, one which combines both the subjective and objective means of gaining knowledge.

AN INTEGRATED SCIENCE OF LIFE

One approach to developing an integrated science of life is to locate certain basic qualities and dynamics of natural law which are common to both the subjective and objective spheres of life.

In our analysis of a unified understanding of natural law from the point of view of modern physics and physiology we have observed a specific trend which suggests that the quality of self-interaction or self-referral may be a fundamental characteristic or marker of the dynamics of evolution. As we have seen, according to the most current theories of physics the universe emerged from a perfectly symmetrical super-field. This super-field, best described by current unified field theory, inherently contains the property of self-referral or self-interaction in its most concentrated form.

The property of self-referral expressed in the process of spontaneous symmetry-breaking is the very basis for the expression of diversity at the inception of the evolution of the universe. As the universe evolves and greater physical diversity is expressed the quality of self-referral becomes less and less evident. It is not until the evolution of more complex physical systems, and ultimately the evolution of physiological systems, that self-referral reemerges. With the evolution of more stable and integrated systems this property of self-interaction becomes increasingly evident in the hierarchy of self regulating homeostatic mechanisms.

If the evolution of life can be seen in terms of self-referral, it might also be associated with other fundamental properties such as coherence. Coherence refers to the degree of correlation between the parts of a system. Systems that display higher degrees of self-referral and self-regulation usually display higher levels of coherence and integration. Thus, for example, the nervous system, with its remarkable ability to coordinate and integrate many seemingly unrelated events, displays a very high level of both self-referral and coherence.

Two other characteristics which seem uniquely related to the property of self-referral are intelligence and consciousness. While these properties at first seem too subjective to consider, they may be viewed independently of what we normally consider to be human intelligence or human consciousness. Consider, for example, the attributes of the unified field as proposed by modern unified field theories. The unified field, according to these theories, is totally self-sufficient and self-interactive. It is also completely unmanifest, existing as a state of perfect symmetry. If we view it as being the most concentrated form of the laws of nature which structure the universe, in the same way that DNA represents a concentrated form of the structure and function of the physiology, then the unified field may be considered a field of pure potentiality or pure intelligence.

Further, since this field has the ability to act on itself or know itself in a completely spontaneous manner, it could be said to display the qualities of pure consciousness or pure awareness. This is precisely the type of argument presented by Hagelin in his comparison of the qualities of the unified field and the subjective state of pure consciousness or pure intelligence as described in the Maharishi Technology of the Unified Field. They are both descriptions of a field which is completely self-sufficient and self-interactive.

As Hagelin points out and we have described above, the evolution of the universe may be seen in terms of an initial state of complete self-referral; this initial state is thus also a state of pure intelligence. It may seem strange at first to assign a level of intelligence or consciousness to a physical system, or even a simple biological system, but when viewed as properties essentially synonymous with the degree of self-referral of the system, then the evaluation of these attributes becomes more meaningful. As we proceed up the scale of biological evolution the entire picture becomes more familiar, each more evolved species displaying greater levels of intelligence and awareness. The highest level of a self-referral system is seen in the human nervous system with its billions of interconnecting neurons and its ability to coordinate all other systems into an integrated whole. This holistic functioning of the individual human physiology gives rise to the extraordinary property of self-awareness in human consciousness.

Vedic Science has for centuries not only recognized these qualities of self-referral, coherence, intelligence and consciousness as characteristic of the unified field of both subjective and objective creation but, more importantly, has prescribed a set of specific neurophysiological procedures to unfold their full expression in each individual. Thus the integration of modern science and Vedic Science provides a complete theoretical understanding of the underlying basis of natural law as well as a practical technology to improve all aspects of life.

The advantages of such an integrated science are numerous and far reaching. From the standpoint of modern science the most basic issue of the relationship between consciousness and matter, which has plagued scientists for several centuries, is suddenly resolved. By identifying the self-referral and self-sufficient nature of the unified field of modern science with the self-referral and self-sufficient nature of the field of pure consciousness in Vedic Science, a common basis for all matter and consciousness is revealed.

In terms of modern neuroscience, this unification between consciousness and unified field theory is particularly relevant, for it is clearly the human nervous system that stands as the interface or link between consciousness and the physical laws of nature. It is the nervous system that, by its different styles of functioning, is able to support and express different states of consciousness. Through the technology of the TM and TM-Sidhi program the human nervous system is able to support and experience the least excited state or pure state of consciousness. Pure consciousness, the unified field of natural law, is a state that is both the ultimate basis of all subjectivity and also the ultimate basis of all objective states. Thus, the nervous system has the extraordinary capacity to directly experience the laws of nature in the state of pure consciousness.

A scientific instrument, such as a particle accelerator, that would allow us to examine the level of the superfield, would essentially have to be as big as the universe. What the Maharishi Technology of the Unified Field provides is a means to experience what material technology cannot possibly probe—to experience the unified field of all the laws of nature within one's own consciousness.

As Maharishi has said:

> The Veda has been declaring throughout time: Amritasya Putrah—O, Sons of Immortality.' From the field of pure knowledge, the mortal has always been welcomed as the descendant of the Immortal. Modern science in its infancy, playing with the fine particles of nature, discovered the destructive potential of natural law, and has delivered total annihilation at the doorstep of human existence. Now it is high time for modern science in its present state of maturity to repay the debt it owes to life. Vedic Science offers the guiding principle. Vedic knowledge is emerging as the most profound science of life, and offers fulfillment to the human quest for perfection. This is the time for modern science to rise to fulfillment.

Chapter 3

THE PHYSIOLOGICAL AND BIOCHEMICAL CORRELATES OF THE LEAST EXCITED STATE OF CONSCIOUSNESS

Advancement in science has always been accompanied by new technologies to probe deeper into nature's functioning. Our current understanding of human physiology has primarily been focused on the study of waking, dreaming and sleep states of consciousness, because of the lack of a technology both to produce and to explore lesser excited states of consciousness.

We have accumulated an extensive body of research on acute states of activation, such as stress, exercise, and physiological changes associated with increased attentional demand. However, because a sufficiently sophisticated technology has been lacking, comparatively little is known of physiological states of decreased activation in which mental and physical activity is minimal yet, unlike sleep or drowsiness, the essential quality of awareness or consciousness is maintained and even heightened.

For thousands of years, such a state of restful inner alertness—referred to as "pure consciousness" or "Samadhi"—has been described not only as a relatively common experience but actually as the goal of various traditions of meditation. Unfortunately, within many of these meditation traditions the specific physical and mental technologies intended to produce this experience have been either inaccessible, impractical, or ineffective. Thus certain key questions concerning the validity of a least excited state of consciousness and the neurophysiological mechanisms which might support it have remained unanswered. With the introduction of the Maharishi Technology of the Unified Field the means to produce this state of pure consciousness have suddenly become available. The result has been an upsurge of scientific investigation at leading institutions around the world.

Objective explorations into various types of meditation techniques have been undertaken in the past. Unfortunately, the primary interest of many of these investigations was peripheral to the study of the nature of consciousness. Their concern was more with the unusual physiological feats that meditating subjects could supposedly perform. Despite this and the fact that many of the physical and mental procedures used were often impractical and difficult, several of these studies do reveal unique physiological changes associated with experiences in meditation, and thus deserve further consideration. In the following section of this chapter we will first briefly review the early research on meditation techniques and then focus on the more recent research concerned with the Maharishi Technology of the Unified Field.

HISTORY OF RESEARCH ON MEDITATION TECHNIQUES

In 1935 a French cardiologist, Therese Brosse, travelled to India and made what appeared to be the first physiological measurements on yogis. In 1947 she reported that one of her subjects was apparently able to stop his heart (Brosse, 1946). It was not until some 20 years later that other researchers monitored subjects practicing various physical and mental yoga techniques (Wenger et al., 1961). In an attempt to follow up and extend Brosse's original study, they measured a number of other physiological variables and found decreased respiratory frequency and increased skin resistance, but no consistent change of heart rate, blood pressure, or EEG during physical and mental yoga exercise. They suggested that the disappearance of the electrical signals from the muscles contracting (as a result of the type of physical maneuver the yogi had performed) obscured the heart impulse signal; this, combined with a limitation in Brosse's equipment, caused the disappearance of the signal of heart activity she had reported.

The conclusions of the authors were that (a) direct control of "involuntary" autonomic functions such as heart rate was probably accomplished through intervening voluntary mechanisms; (b) yogic meditation represented deep relaxation of the autonomic nervous system without drowsiness or sleep; (c) there were possible beneficial applications; and d) further research was necessary (Bagchi and Wenger, 1957; Wenger and Bagchi, 1961).

The authors also commented on the problems in obtaining and selecting expert subjects. They were primarily interested in individuals who claimed they could control their autonomic functions (heart, breath rate, blood pressure, etc.). Due to the authors' interest in phenomenal changes of self-control (such as extreme decrease of heart rate), many of the subjects they chose to study primarily utilized techniques of physical control; and it is possible that few were actually experts in meditation.

The problem of subject selection and measurement in a conventional and consistent manner is found in most research on yoga. Today there are many techniques that are referred to to as "yoga"; most of the research has investigated rather difficult and awkward techniques that demand deep concentration and impractical conditions. Perhaps for these reasons results are often inconsistent. Studies of oxygen consumption illustrate some of these points. Independent studies by two researchers, Rao (1962) and Miles (1964), report significant increases in oxygen consumption during various physical yoga exercises while three other teams of researchers (Hoenig, 1968;

Karambelkar et al., 1969; Rao, 1968) suggest a decrease of oxygen consumption during yoga meditation techniques.

Another problem is that in some studies on yogic techniques rather unconventional methodologies have been used. For example, in one study the subject was buried in an "air tight pit" and was found comatose after 72 hours (Vakil, 1950). Needless to say, the unconventional nature of the procedures and wide variety of techniques studied leave a great deal to be desired as far as scientific procedure and credibility is concerned.

In the area of electroencephalography or brain wave research, early studies utilized more reliable procedures. Since its discovery by Hans Berger (1929), the surface recording of human brain wave patterns, known as the electroencephalogram (EEG) has been a widely utilized tool to study consciousness. Today the EEG patterns are broadly classified, according to the frequency of the electrical wave, into the following categories (see table 1).

Table 1. Classification of EEG Patterns

Type of Wave	*Frequency*	*Condition When Present*
beta	18-30	awake, active, eyes open
alpha	8-12	awake, relaxed, eyes closed
theta	5-7	awake, in children
delta	.5-4	asleep

In one of the first pilot studies on EEG during meditation two French researchers, Das and Gastaut (1957), studied EEG in a yoga practitioner. They reported the occurrence of so-called alpha waves during meditation. During what they described as a deeper state of meditation these alpha waves gave way to periods of faster beta and gamma waves. Anand and co-workers (1961) studied four yoga practitioners and reported prominent alpha-wave activity before and during the practice of their meditation, with an increase in alpha-wave amplitude during meditation.

Some of the most interesting and extensive early studies on meditation states were conducted in Japan on Zen meditators. Kasumatsu and Hirai (1966) reported that the Zen meditators showed abundant alpha wave activity, even with their eyes half open. During their meditation, the alpha wave also increased in amplitude and regularity especially in the frontal and central areas of the brain. The more experienced Zen monks showed such changes as significant decreases in oxygen consumption, respiration rate, and slight increases in pulse rate and blood pH during meditation (Akishige, 1968; Sugi and Akutsu, 1968).

Taken together, the consistent changes seen in Zen meditators and in several of the yoga meditators suggested a unique physiological state. However, especially in the studies on various yoga practices, several problems, such as a lack of a systematic meditation procedure among subjects and interest in manipulative procedures rather than genuine meditation techniques, have limited the studies.

EARLY STUDIES ON THE TRANSCENDENTAL MEDITATION TECHNIQUE—A PROPOSED FOURTH MAJOR STATE OF CONSCIOUSNESS

For scientists, the introduction of the Transcendental Meditation technique by Maharishi offered a great opportunity to objectively measure the physiological, biochemical, and psychological correlates of meditation. From the moment Maharishi first introduced the TM technique into the West in the early 1960's, he encouraged research in all areas. In England and Germany some preliminary work was undertaken. However, no major research study actually began until the late 60's, perhaps because of a still pervasive atmosphere of uncertainty and lack of interest in meditation procedures among the scientific community. For my doctoral thesis in the Department of Physiology at the University of California at Los Angeles, I had the opportunity to conduct a comprehensive study entitled "The Physiological Effects of the Transcendental Meditation Technique—A Proposed Fourth Major State of Consciousness" (Wallace, 1970a). The work was conducted in collaboration with and under the supervision of several excellent researchers, particularly Drs. Archie Wilson and Donald Walter, then with the *Department of Physiology* and the *Brain Research Institute* at UCLA.

To a large extent, Maharishi had predicted the basic types of changes that were to be found in the early studies and in more elaborate studies to follow. For example, in his book *The Science of Being and the Art of Living*, published in 1963, he wrote:

> When the mind transcends during Transcendental Meditation, the metabolism reaches its lowest point; so does the process of breathing; and the nervous system gains a state of restful alertness which, on the physical level, corresponds to the state of bliss consciousness or transcendent Being.

In other sections of the book, Maharishi describes this state of restful alertness as a fourth major state of consciousness distinct from waking, dreaming, and sleeping. The first major publication to verify Maharishi's predictions occurred in 1970 in the journal *Science* in an article entitled "Physiological Effects of the Transcendental Meditation Technique" (Wallace, 1970b). This early research, conducted by myself and coworkers, was extended and published in several other journals (Wallace et al, 1971,

Wallace et al, 1972). Very briefly, since a more detailed description of these findings is given later, the results of the studies were:

(a) deep rest as indicated by a marked and significant decrease in oxygen consumption and carbon dioxide elimination,

(b) significant decreases in respiration rate, minute ventilation, and heart rate,

(c) deep relaxation as indicated by a significant and sharp increase in skin resistance,

(d) normal maintenance of critical physiological functions as indicated by stable arterial levels of partial pressure of oxygen and carbon dioxide, pH and blood pressure (blood pressures were quite low throughout the experiments),

(e) significant decrease in arterial blood lactate and

(f) restful alertness, as indicated by EEG changes showing an increase and spreading of alpha and theta wave activity to the more central and frontal areas of the brain.

The overall conclusion of the studies was that Transcendental Meditation produces a unique state of restful alertness, indicative of a fourth major state of consciousness that is physiologically and biochemically unique. Since these initial findings, many hundreds of studies have been completed in all fields which, when taken together, support and enrich the concept of a transcendental or fourth state of consciousness.

Later research has helped clarify several important

issues that were not clearly defined in the earlier research, perhaps the most important being the identification of the state of pure consciousness itself. During the practice of the TM technique there appears to be a mixture of states; the technique itself is a dynamic one having both an "inward phase" and an "outward phase." In the inward phase subjects report their mental activity settling down to quieter levels until they eventually transcend all mental activity and reach a ground or least excited state. In the outward phase they report emerging from the state of pure consciousness and the gradual appearance of more excited states of mental activity, until they begin the inward phase once again. The degree of clarity of the experience of pure consciousness subjectively reported varies greatly, as does the frequency and duration of each of these inward and outward phases. Subjects who are overly tired before practicing the TM technique report having short periods of drowsiness or even sleep during meditation. Through the entire 20 or 30 minutes of the TM technique there is often a mixture of active waking state, drowsiness, sleep or relaxation, quiet waking state, and transcending to a least excited state of pure consciousness.

According to Maharishi, if the nervous system were free from any functional or structural abnormalities, then the state of pure consciousness would be maintained for long periods. This is confirmed by other studies on advanced practitioners of the TM and TM-Sidhi program which we will examine later. If however, some abnormalities in the physical system exist, as the nervous system begins to settle down to the least excited state it automatically normalizes itself; it causes itself either to rest through drowsiness or sleep, or to return to a more excited state. The mixture of states that occurs during the practice of the TM technique thus is a direct consequence of the dynamics of the procedure and the initial condition of the individual's nervous system. That the earlier research, which we could refer to as phase I studies, did not carefully discriminate between this mixture of states during meditation, has to some extent led to confusion over the difference between the physiological correlates of pure consciousness and those of relaxation and drowsiness. Later studies, which we could refer to as phase II studies, have taken care to selectively study periods of pure consciousness, and as we shall see, have demonstrated far more marked and unique physiological changes.

Another difficulty that has sometimes arisen in the attempt to discriminate between the state of pure consciousness and other states is the tendency of researchers to use physiological parameters that are too few and too limited. A classic example of the problems of using limited measurements can be seen in the first attempts to distinguish between dreaming and waking through the use of EEG patterns. The EEG pattern during dreaming looks extremely similar to that during waking, and yet behaviourally the states are entirely different. Based on these EEG patterns dreaming was initially referred to as "paradoxical" sleep. It was not until the observation by Aserinsky and Kleitman (1965) of so-called rapid eye movements (REM) and the subsequent recording of electrocculograms (EOG) and electromyograms (EMG) during dreaming that clearly defined criteria were established which enabled physiological discrimination between dreaming, sleep and wakefulness. Researchers have further characterized dreaming by variations in other physiological parameters (Dement and Masserman, 1964; Jouvet, 1962; Oswald, 1962; Synder et al., 1964).

The same point applies to research on the TM program. If only one or a few parameters are used it may not be possible to discriminate among the various mixtures of states that occur during the TM technique. In early research a number of parameters were used in an attempt to undertake a multidisciplinary approach to the study of consciousness. Subsequent research has demonstrated, however, that there exist other, more discriminating, parameters which are able to better characterize the state of pure consciousness. Unless these measures are utilized, the unique characteristics of this state may remain obscured. In the following sections we will consider various studies conducted on the TM technique in the light of these methodological issues.

METABOLISM AND RESPIRATION

The metabolic rate is perhaps the most fundamental measurement of the overall level of activity or rest in a system. In almost all animals the main metabolic pathways responsible for the vital processes of life are aerobic, that is, they depend upon the presence of oxygen. Since oxygen is such a critical component of aerobic metabolism, its measurement is most commonly used as an indirect means of assessing the metabolic rate of the body. There are a number of standard procedures for the measurement of oxygen consumption, the most reliable being the so-called open circuit method (Benedict and Benedict, 1933).

In the first studies on the TM technique at UCLA, oxygen consumption and carbon dioxide elimination were measured in 20 subjects—a closed circuit method was used for the first 5 and the more reliable open circuit method for the remaining 15 subjects. We reported (Wallace et al., 1971) a marked decrease in oxygen consumption of about 40 cc/min., or 16% during the meditation period as compared to the premeditation period (each subject acting as his own control). The carbon dioxide elimination decreased about 30 cc/min. while the respiratory quotient (the ratio of the volume of carbon dioxide eliminated over the oxygen consumed) remained unchanged. Respiration rate and minute ventilation also decreased significantly without significant change of arterial P_{O_2} and P_{CO_2}. However, arterial pH declined slightly due to a mild but significant metabolic acidosis.

Change in Metabolic Rate

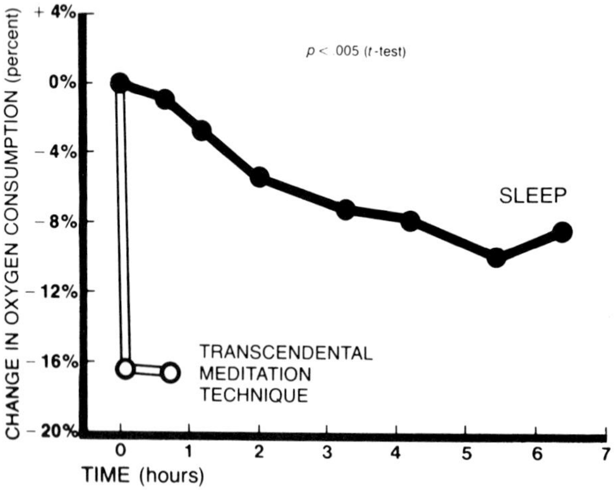

Finding: Metabolic rate was obtained by measuring oxygen consumption in 20 subjects during the Transcendental Meditation technique. The study showed that during the TM technique oxygen consumption markedly decreases. The average decrease was 16 percent within the first ten minutes of a session of the Transcendental Meditation technique. Further, the study showed that the partial pressures of oxygen and carbon dioxide in the blood remained essentially constant. The decrease in metabolic rate during the Transcendental Meditation technique was deeper and much quicker than during deep sleep.

First Reference: Robert Keith Wallace, "The Physiological Effects of Transcendental Meditation: A Proposed Fourth Major State of Consciousness" (Ph.D. Thesis, Department of Physiology, University of California, Los Angeles, California, U.S.A., 1970).

Second Reference: Robert Keith Wallace, Herbert Benson, and Archie F. Wilson, "A Wakeful Hypometabolic Physiologic State," *American Journal of Physiology* 221, no. 3 (U.S.A.: 1971): 795-799.

These and electrophysiological measurements (reviewed in detail below) suggested that the TM technique produced a wakeful hypometabolic state with decreased sympathetic activity which is physiologically distinct from ordinary waking and sleep.

Seven subsequent studies have confirmed these initial findings. Studies done by Corey (1973) on 20 subjects replicated the early findings at UCLA, showing a decrease in oxygen consumption of approximately 20%, and further showing a decrease in airway resistance. Dhanaraj and Singh (1973) also showed a similar, significant decrease—approximately 15%—in oxygen consumption in subjects practicing the TM technique, while nonmeditating relaxing controls showed no significant change. Likewise Reddy (1976) showed a similar decrease in oxygen consumption and carbon dioxide elimination in TM subjects as compared to controls. One further study by Farrel (1979) showed a significant decrease in oxygen consumption with no change in respiratory quotient in nineteen subjects practicing the TM technique.

In addition, Throll (1982) measured oxygen consumption, total volume, respiration rate, heart rate, and systolic and diastolic blood pressure in 39 subjects learning the TM technique, and in 21 subjects learning Jacobson's Progressive Relaxation. Subjects were tested both during the techniques and also on a longitudinal basis, before learning either technique, immediately after, and at five, ten and fifteen weeks later. There were no significant differences between groups for any of the variables at pretest. However, the TM group displayed more significant decreases during meditation than the progressive relaxation group in respiration rate, oxygen consumption, and total volume. The TM group also showed significant long-term reductions (as compared to the progressive relaxation group) in the following parameters: heart rate, respiration rate, total volume and diastolic blood pressure.

Gallois (1984) compared 10 subjects practicing the TM technique with 10 subjects practicing autogenic training and 10 control subjects. The results of the study showed a marked decrease in respiration rate in the TM group as compared to the other groups, with the appearance of frequent respiratory suspensions in the TM group, reaching a maximum duration of 50 seconds. In addition, simple reaction time decreased slightly in the TM group, where as it increased slightly in controls. Finally, Garnier and coworkers (1984) studied pulmonary ventilation before, during and after the TM technique as compared to relaxing controls. The oxygen uptake per kilo weight of the TM subjects was reported to be 15.6% below that of the control group during meditation. This value decreased by a further 8.5%, indicating that the TM subjects showed values below their calculated basal metabolic rate.

The results of certain of these studies suggest the importance of looking at continuous measurements so that the "deeper" periods of meditation can be identified, and, further, of studying more advanced subjects who report clear experiences of pure consciousness. To some extent, this was done in one of the first studies on the TM technique. Allison (1970), in a report on one advanced subject, found that respiration rate decreased from 12 breaths/min. before meditation, to 4 breaths/min. during, and again returned to 12 breaths/min. after meditation, with no evidence of compensatory breathing.

A recent more extensive study by Farrow and Hebert (1982) has helped clarify these previous results. This study analysed specific periods during the TM technique that more carefully isolates and studies the experience of the least excited state of consciousness. In four independent experiments Farrow and Hebert attempted to determine frequencies of breath-stoppage episodes during the TM technique and whether these episodes were correlated with the experience of pure consciousness.

Patterns of Respiratory Suspension during the TM Technique

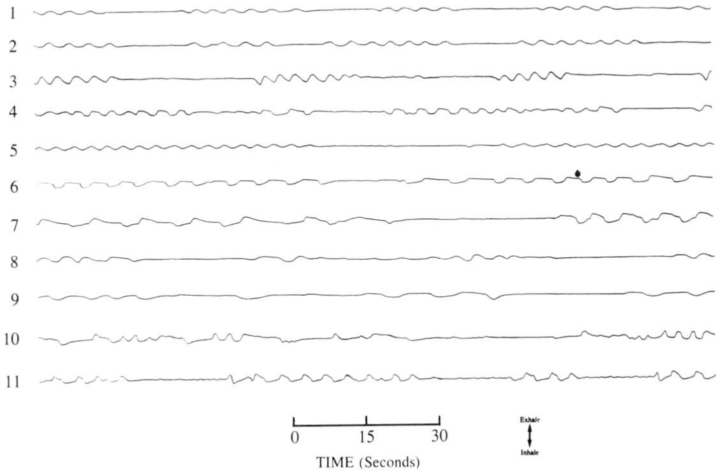

REPRESENTATIVE SECTIONS OF RESPIRATION TRACINGS IN OUR SUBJECTS (NOS. 1–10) SHOWING PERIODS OF RESPIRATORY SUSPENSION MEETING OUR CRITERIA. .

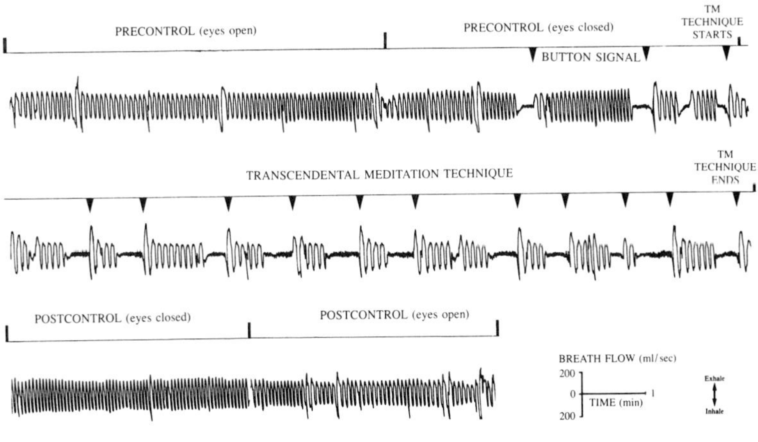

PERIODS OF TRANSCENDENTAL CONSCIOUSNESS DURING THE TRANSCENDENTAL MEDITATION TECHNIQUE. The subject was asked to press an event-marker button after each subjective experience of transcendental consciousness (upper trace in each set). Each mark the subject made closely corresponds to the end of a period of markedly reduced breath flow (lower trace in each set).

Farrow, J. T., and Herbert, J. R. Breath suspension during the Transcendental Meditation technique. *Psychosomatic Medicine*, 44(2): 133–153, 1982.

In the first experiment, 95 subjects who had been practicing the TM technique from one month to 13 years were studied before, during, and after meditation. Eleven of the subjects exhibited a total of 151 breath suspension episodes, almost all occurring during the TM technique. In this experiment respiration was considered suspended if for ten seconds or more the pen tracing did not fluctuate significantly. In the second and third experiments the following improvements were made: 1) the criteria were altered so that they accounted for individual variations in average respiration time (respiratory suspension was noted if the pen tracing did not fluctuate significantly for two respiratory cycles); 2) more advanced TM subjects (some practicing the TM-Sidhi program) were studied as well as nonmeditating controls; and 3) the measuring equipment was less intrusive. As a result, the frequency of breath suspension episodes was greater. For example, in the second experiment, 21 of the 28 TM subjects exhibited a total of 116 episodes while 9 of the 23 controls exhibited only a total of 14 breath suspensions. The frequency of episodes and mean, maximum, and total episode lengths were all substantially greater in the TM group than in the control group.

In experiment three there was an attempt to correlate the occurrence of breath suspensions with the subjective experience of pure consciousness. In this experiment all subjects had participated in advanced courses designed to deepen and extend the clarity of experience in meditation; all reported frequent and sustained experiences of pure consciousness during meditation. Subjects were given an event marker button and instructed to push the button after each experience of pure consciousness. Respiration was measured in a nonintrusive manner (using a two-channel paramagnotometer). Eight of 11 subjects exhibited 57 periods of breath suspension. A large percentage of the button presses noting the experience of pure consciousness, occurred within 10 seconds of the offset of one of the 57 breath suspensions. The temporal distribution of button presses was significantly related to the distribution of breath suspension episodes, indicating that breath suspension is a physiological correlate of some, but not all, episodes of pure consciousness.

In experiment four one very advanced TM meditator, who reported frequent clear experience of pure consciousness, was studied over six sessions. Respiration rate, minute ventilation, oxygen consumption, basal and phasic skin resistance, heart rate, electrocculogram (EOG), and electroencephalogram (EEG) were measured. In three sessions the subject was given an event marker button to press after each experience of pure consciousness. In almost all cases the event marks occurred immediately after a period during which breath flow decreased to nearly zero, indicating a clear and consistent correlation between periods of respiratory suspension and the subjective experience of pure consciousness. Event marks and periods of breath suspension were not present in the records during either the precontrol, eyes-open period or the postcontrol, eyes-closed, and eyes-open periods; however, some did occur near the end of the precontrol, eyes-closed period. As in the case with many subjects, there is often a spontaneous drifting into the practice of the TM technique during the precontrol eyes-closed period.

Over five sessions, there were 191 breath suspensions, occurring at a mean rate of one every 52 sec. The mean period of breath suspension and experience of pure consciousness was about 18 seconds, with the longest period about 35 seconds. A detailed analysis of the air flow during periods of breath suspension showed that it did not entirely stop but continued with high frequency, low amplitude fluctuations at about 2Hz and 4.5Hz.

Two key questions asked by Farrow and Hebert in this experiment were whether or not the breath suspension periods were intentional and whether they indeed reflected a true reduction in the metabolic needs of the body. Concerning the first question, in all experiments subjects were blind to the purpose of the experiment; several sessions employed further equipment and conditions which reduced or eliminated the subject's awareness that respiration was being measured. In a number of the sessions, multiple electrodes were attached for other measurements, thus further lessening the subjects' awareness of the purpose of the experiment. To determine whether the breath suspensions were intentional, the compensatory hyperventilation seen after intentional breath suspension periods was analysed and compared to breathing activity following respiratory suspensions during the TM technique. A subject asked to intentionally hold his breath for time periods similar to the TM respiratory suspensions showed a significant and marked increase in minute ventilation (2.71 liters per minute), as compared to a nonsignificant increase (0.57 liters per minute) after the periods of respiratory suspension during TM meditation.

Another concern was to distinguish the periods of respiratory suspension seen during sleep apnea from those during the TM technique. This was determined by the fact that EOG measurements during the TM technique did not show the typical, slow-rolling eye movements preceding or associated with sleep apnea.

Several measurements were made to answer Farrow and Hebert's second question: whether the periods of respiratory suspension represented a true reduction of the metabolic needs of the body. Measurements of respiration rate during the TM technique as compared to the mean of the pre and postcontrol periods showed a 50% decrease, while measurements of minute ventilation showed a significant decrease—approximately 35%—during meditation. Oxygen consumption was monitored in one session and was found to decrease a mean of 40% during the TM period, reaching a maximum of 60%.

However, because the technique used to measure oxygen consumption was novel, the researchers felt the need for further replications.

These findings, while not completely answering the question of whether these periods represent a dramatic decrease in metabolic requirements do clearly distinguish between the experience of pure consciousness during the TM technique and mere relaxation during rest or some specific relaxation procedure. The patterns of respiratory suspensions which accompany the subjective experience of pure consciousness are different from those seen in other states. By using advanced subjects with clear experiences of this ground state of consciousness and by attempting to isolate the experience of pure consciousness rather than averaging over a mixture of states, a much more detailed analysis was achieved. A further description of the other physiological parameters, such as skin resistance and EEG power and coherence, utilized by Farrow and Hebert to characterize the periods of pure consciousness, will be discussed in a later section along with one other study which includes a comprehensive analysis of EEG coherence changes during periods of respiratory suspension.

Several other recent studies have extended Farrow and Hebert's results and have more carefully analyzed the neurophysiological control of respiratory patterns during the TM technique. For example, Wolkove and co-workers at McGill University in Canada, comparing long-term TM subjects with nonmeditating controls at rest, confirmed Farrow and Hebert's findings of both a significant decline of minute ventilation and the observation of periods of respiratory suspension in TM subjects. Further these researchers along with another researcher in Australia both found that when TM subjects were given increasing amounts of carbon dioxide in inspired air there was a reduced respiratory response. This suggested a decreased sensitivity to high carbon dioxide concentration and an alteration in the neurophysiological mechanisms which regulate breathing (Wolkove et al., 1984; Singh, 1984).

The most extensive investigation in this area is that of Kesterson (1986). In this study a cross section of groups of TM and TM-Sidhi participants were tested during their group practice of the TM-Sidhi program. Three categories of subjects were identified according to their pattern of breathing while practicing the TM technique.

The first group showed no changes in the frequency of breathing with meditation, the second a large decrease in the rate of breathing, and the third group prominent, frequent periods of respiratory suspension. The subjects in the third group were studied extensively in order to determine the underlying neurophysiological mechanisms producing the suspensions. Similar to Wolkove and coworkers, and Singh findings, these subjects demonstrated decreased sensitivity to high levels of carbon dioxide in ambient air during meditation. Further, they showed an increased sensitivity to low levels of oxygen. It was also discovered that many of the subjects demonstrating spontaneous suspensions during meditation were, in fact, breathing apneustically, i.e. the suspension began with a full inspiration and continued with a gradual inspiration of between 100 to 300 ml/min during the suspension. He theorized that the decreased sensitivity to carbon dioxide that occurs with the onset of meditation (probably due to an inhibition of the off-switch mechanism controlling inspiration) results in a decrease in the frequency of breathing which, in turn, causes an increase in the carbon dioxide in the blood. The increase in carbon dioxide in the blood changes the sensitivity of the carotid bodies to the lack of oxygen such that they begin to fire more rapidly at higher partial pressures of oxygen. The increased sensitivity to lack of oxygen then causes the termination of the prolonged inspiration (apneusis) and the initiation of expiration (normally it is the buildup of carbon dioxide that controls the onset of expiration).

Interestingly, the carotid bodies project to both the respiratory nuclei and the reticular activating system (RAS), which is partly responsible for "wakefulness" or arousal. At the end of a suspension these subjects not only begin to breath again, but are aroused from the experience of transcendence by the increased afferent traffic from the carotid body to the RAS, thus explaining the high correlation of button presses and suspensions as reported by Farrow and Hebert (1984). The subjects are not aware of the state of transcendental consciousness while they are experiencing it, but realize that they were absorbed in it upon arousal.

Furthermore, Kesterson discovered a drop in the respiratory quotient (RQ—the ratio of carbon dioxide produced/oxygen consumed) for almost all subjects in his experiment during meditation. He hypothesized that this drop was a consequence of mild hypoventilation, which is known to decrease RQ. He demonstrated in a series of separate experiments that, in fact, alveolar ventilation decreased significantly greater than oxygen consumption for meditators but not for non-meditating controls while relaxing, an argument in favor of hypoventilation. He cites this change as further evidence for brainstem inhibition of respiratory control during TM.

Kesterson suggests his findings are contrary to previous models hypothesizing that the respiratory changes observed during TM originate in decreased metabolic needs. His findings indicate that they are a result of specific alterations in neurophysiological centers within the brain, more specifically those respiratory centers in the medulla involved with the regulation of inspiration and expiration. Kesterson's research, as well as several other studies suggests that oxygen consumption may not be as good a discriminator of transcending as respiratory patterns. Oxygen consumption is sensitive to bodily movements and any state of rest or relaxation will result in a gradual decrease in metabolic rate. The specific patterns of respiratory activity during the TM technique seems to be better indicators of the subjective experience of transcending, expecially when coupled with other physiological measurements such as EEG coherence.

The most important conclusions drawn by Kesterson, however, are those concerning changes in states of consciousness. Sullivan (1980) has noted that respiratory pattern is a good discriminator between sleep, dreaming and wakefulness, i.e. states of consciousness. In fact, he suggests that if respiratory physiologists had discovered REM sleep instead of EEG researchers, it would have been called RERM, Rapid Erratic Respiratory Movements. Deep sleep is characterized by slow, periodic monotonous patterns, REM sleep by irregular variations unrelated to carbon dioxide control, and wakefulness provides a major component of respiratory drive during daily activity. Brainstem inhibition of respiratory centers, a buildup of carbon dioxide, a drop in RQ and mild hypoventilation are all known to accompany the transition between these states of consciousness. The respiratory patterns demonstrated by advanced TM meditators, and investigated by Kesterson, are indicative of a transition in the state of consciousness and a state of restful alertness. The subjects who showed the greatest changes in breathing pattern while meditating were also the most alert and reported the best experiences. Kesterson argues that the changes in breathing pattern signify an integration of the three separate states of consciousness into a single, unified state.

LACTATE GENERATION

One of the first biochemical findings noted in subjects during the TM technique was a marked decrease in the level of lactate in arterial blood and a continued low level afterward (Wallace et al., 1971). Lactate or lactic acid is the byproduct of a less efficient type of metabolism known as anaerobic metabolism, which occurs when cells

Change in Blood Lactate

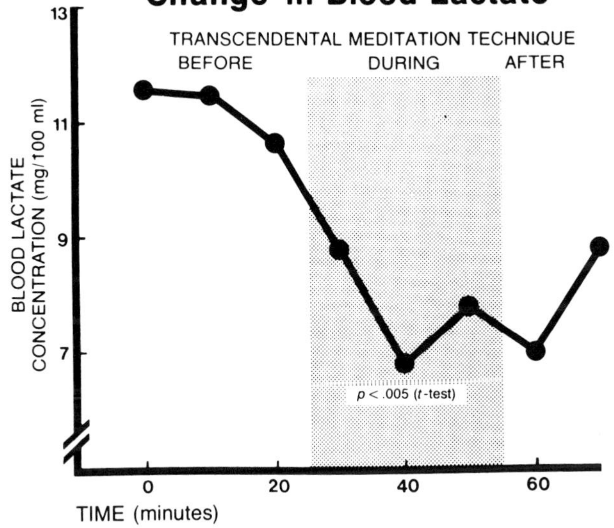

Finding: During the Transcendental Meditation technique the concentration of blood lactate markedly decreases and remains low some time after practice of the technique.

First Reference: Robert Keith Wallace and Herbert Benson, "The Physiology of Meditation," *Scientific American* 226, no. 2 (U.S.A.: 1972): 84–90.

Second Reference: Robert Keith Wallace, Herbert Benson, and Archie F. Wilson, "A Wakeful Hypometabolic Physiologic State," *American Journal of Physiology* 221, no. 3 (U.S.A.: 1971): 795–799.

are not able to utilize oxygen. Even under normal conditions anaerobic metabolism exists and, as a result, a certain amount of lactate is constantly being produced by various cells in the body. Lactate production is dramatically increased in situations where the oxygen supply is eliminated or decreased in certain cells in the body—for example, in the skeletal muscle cells of a runner who is sprinting the last hundred yards of a mile race. The cells must rely on anaerobic metabolism even though it is less efficient. During these last hundred yards the runner is building up what is known as an oxygen debt, which is indicated by the increased production of lactic acid or lactate. If too much builds up too quickly, an excessive acidic condition can be created in the body, which can be detrimental to health.

At the University of California at Irvine Drs. Jevning and Wilson replicated and extended the early findings on decreased levels of arterial lactate during the TM technique in a series of well-controlled experiments, comparing TM meditators to nonmeditating controls (Jevning et al., 1978a, 1983a). Perhaps the most interesting of these experiments involved an in vitro study on lactate generation in arterial red blood cells during the TM technique (Jevning et al., 1983a).

Meditators and a nonmeditating control group were studied in 40-minute periods before, during, and after meditation or simple relaxation. Samples of arterial blood were drawn during each of these periods incubated for 90 minutes at 37 degrees C to determine rate of lactate generation. The nonmeditating control group showed no significant changes while the TM subjects showed a marked decrease in lactate generation. The samples were also analyzed for pH, P_{O_2} and P_{CO_2} and hematocrit. No significant alteration in these parameters was seen for either the meditation or relaxation periods. The authors concluded that the TM technique is associated with the inhibition of red blood cell glycolysis and hypothesized the possible production of a plasma factor, which is not one of the ordinary factors involved in lactate generation during glycolysis. Further research in this area is continuing at Drs. Jevning and Wilson's laboratory, supported by research grants awarded from the National Institutes of Health. Jevning and Wilson feel their results suggest that an unknown chemical factor, which affects basic metabolic processes, is being produced and released into the blood as a result of the TM technique.

REDISTRIBUTION OF BLOOD FLOW AND MUSCLE METABOLISM

A number of circulatory changes have been reported in subjects practicing the TM technique as compared to nonmeditating controls. In one study, hepatic blood flow was reported to decrease 40% in TM subjects, but not at all in controls; cardiac output increased 15% in TM subjects, but not at all in controls; and renal blood flow decreased 20% in both TM subjects and controls (Jevning et al., 1978a). Jevning and coworkers concluded that, since arterial blood pressure has been reported to be constant (Wallace et al., 1971), and since there is a marked reduction in hepatic and renal blood flow, there must be an increase in blood flow to some other major organs than the heart, namely the muscles or brain.

Earlier experiments examined forearm muscle blood flow using a water plethysmograph, and found a small but significant increase in blood flow during the TM technique as compared to a pre-test period (Levander et al., 1972). The same subjects, when tested on different days and asked to sit quietly and not meditate during the experiment, showed slight decreases in blood flow during the test and post-test periods. Jevning, Wilson and coworkers (1983b), using a different type of technique, also found a significant but small increase in forearm muscle blood flow during the practice. According to their calculations, this increase in forearm muscle blood flow was not nearly enough to account for the postulated increase in total blood flow. Jevning and coworkers therefore reasoned that there might be an increase in brain blood flow. Using a noninvasive impedance technique (rheoencephalography), they demonstrated a significant and marked increase in brain blood flow (Jevning et al., 1978b).

Jevning and Wilson's further studies of forearm muscle blood flow reveal another intriguing and unexpected finding. In addition to measuring blood flow they measured the partial pressure and absolute levels of oxygen and carbon dioxide of the arterial and venous blood in the forearm of TM subjects during meditation and in nonmeditating controls during rest. From these measurements they were able to calculate the metabolic activity of the forearm muscles. They found the normal difference in oxygen content of the arterial and venous blood was significantly reduced in TM meditators during meditation, as compared to controls during rest. This suggest that less oxygen was being used by the cells, indicating a state of lower metabolism and deeper rest in the muscles of the TM subjects.

More interestingly, they found the normal difference in carbon dioxide content of arterial and venous blood was markedly reduced in TM subjects during meditation, with a less marked reduction in the controls during rest. These findings indicated a decreased or complete cessation of carbon dioxide production by the forearm muscles after 20-30 minutes of TM. Calculations of the respiratory quotient during meditation showed a value of almost zero, as compared to the normal value of about .60. Thus a dramatic alteration in local metabolism occurs as a result of the TM technique. This change is particularly interesting in the light of the earlier research, which showed no changes in the systemic levels of oxygen and carbon dioxide as reflected by arterial P_{O_2} and P_{CO_2} (Wallace, et al., 1971).

Wilson (1984) speculates that the changes may be due to an alteration in fat metabolism. In order for carbon dioxide not to be produced, it is necessary to postulate that the two or three carbon fragments produced do not enter the tricarboxylic acid cycle. Two metabolic pathways which may not involve carbon dioxide production are beta-oxidation of fatty acids and glycolysis. Wilson and coworkers suggest that fatty acid oxidation is the more likely of these pathways for a number of reasons: one is that the previously observed decrease in blood lactate during the TM technique suggests that the end products of glycolysis are not being formed. In order to account for the low arterial-venous carbon dioxide content is also necessary to explain why the products of fatty acid metabolism do not enter the tricarboxylic acid cycle. Wilson and coworkers speculate that under the resting conditions of the experiment the energy production may exceed utilization, producing conditions which inhibit the further normal metabolism of these products. These unmetabolized products must undergo further oxidation at some other site or organ in the body. The reported marked increased in brain blood flow suggests that perhaps oxidation of these products is occurring there. Further research is being conducted in this area in order to reveal the mechanics and significance of these findings.

BIOCHEMICAL AND HORMONAL CHANGES

Several different research groups have investigated biochemical and hormonal changes during the TM technique. In a series of experiments by Jevning, Wilson and co-workers (Jevning et al., 1977; Jevning et al., 1978c, 1978d) plasma levels of cortisol, prolactin, testosterone growth hormone, and thirteen neutral and acidic amino acids were studied before, during, and after practice of the TM technique. The design of these experiments involved three groups consisting of: 1) a nonmeditating relaxation group of subjects who volunteered for the study prior to learning the TM technique, 2) the same group restudied after four months of TM practice, and 3) long-term regular participants in the TM program. During the experimental period the controls were asked to relax maximally without specific instructions; those practicing the TM technique were asked to meditate. All subjects had been fasting for 12 hours, and an arterial catheter had been inserted into their brachial artery at least two hours before the experiment began, thus allowing for hormonal levels to normalize. Samples were then drawn every 20 minutes throughout the experiment. Measurements were taken in the afternoon when circadian changes, especially in cortisol, are relatively stable.

Plasma cortisol level decreased by a small but significant amount in long-term TM meditators during the meditation period, as compared to the pre and post experimental control periods. Plasma cortisol levels did not change in the nonmeditating relaxation group. After this group was taught the TM technique and retested three to four months later, cortisol levels decreased but did not reach statistical significance. No significant change in growth hormone and testosterone concentration was found in any of the groups. Meditators did not, however, exhibit a post-experimental decrease in testosterone, which was seen in the nonmeditating controls after relaxation.

Prolactin concentration remained stable during meditation but increased significantly afterwards. The nonmeditating relaxation control group showed no such effect. When this group was later taught the TM technique and retested, it also showed an increase in prolactin in the post meditation period.

Of the thirteen neutral and acidic amino acid levels studied only phenylalanine was shown to change. Phenylalanine concentration increased significantly (20%) during meditation. This change was seen only in long-term meditators and not in the nonmeditating group or the restudied controls.

Jevning indicates that these changes, particularly the decrease in plasma cortisol, are in agreement with other physiological findings such as decreased oxygen consumption, heart rate, and breath rate. Jevning further suggests that, since there is a close qualitative relationship between ACTH and cortisol, the decline of plasma cortisol during and after the TM technique reflects decreased pituitary-adrenal activation, rather than a nonspecific increase in the metabolic clearance of cortisol. The finding of decreased blood flow to the liver, the principal site of cortisol degradation, further supports this hypothesis. Concerning the significant increase in plasma levels of phenylalanine in long-term meditators as compared to controls, the authors suggest that since phenylalanine is a precursor to the synthesis of a number of important brain neurotransmitters, such as adrenalin and dopamine, its increase in the blood during TM may indicate an alteration in the utilization of these neurotransmitters by the brain.

Jevning further interprets the combined changes in cortisol, plasma phenylalanine, prolactin, and blood flow as distinguishing the TM technique from other hypometabolic states such as sleep. He notes that in slow wave sleep, for example, there are no acute changes in plasma cortisol and plasma phenylalanine, while other amino acids may decline. The changes in prolactin are particularly interesting since they suggest possible changes in the neurotransmitter dopamine. Dopamine inhibits secretion of prolactin, and changes in levels of prolactin have been used by some researchers as an index of dopamine activity.

Bevan et al. (1976,1980), in an extensive study of both long and short-term changes in TM participants, measured urinary catecholamine levels (UCA), plasma and urinary free cortisol (UFC), serum thyroxine (T4), and triiodothyronine (T3). Three separate studies were performed on normal subjects to examine the possible endocrine effects of various relaxation procedures. First, chronic endocrine effects were investigated in subjects practicing the TM technique, progressive relaxation, autohypnosis, and yoga meditation, both before training and at regular intervals over eight months during training. A control group that was given no specific relaxation instruction, and a group of experienced (more than one year of practice) TM subjects were studied concurrently. Second, acute endocrine effects were studied in all groups by serial blood sampling before and after a half-hour mid-afternoon relaxation session on a Sunday. Urine samples were collected over the same weekend and the following Monday. The third study was conducted during a TM weekend residence course on subjects practicing meditation for longer periods of time. Both experienced and novice TM subjects were investigated.

The results of the first study showed that during the eight months no trends in endocrine levels relating to the relaxation technique practices were apparent for all groups. Overall, however, urinary free cortisol levels of experienced TM subjects were significantly lower than those of novice TM subjects. No significant differences were seen between urinary free cortisol levels of novice meditators and controls, and no significant group differences in urinary catecholamines were found.

In the second study, a highly significant decrease in plasma cortisol following meditation was observed in the experienced TM group. No significant plasma cortisol changes were found in any of the other groups. Levels of T4 fell immediately following meditation in the experienced TM group; however, no significant changes in T3 were observed. Analysis of urinary free cortisol and urinary catecholamine showed a general decline from Saturday to Sunday and an increase on the following Monday, although these trends were not significant.

In the third study, urinary free cortisol levels of

experienced TM subjects fell significantly from the Saturday to the Sunday of the TM residence course, and levels remained lower on the following Monday. Novice meditators showed a similar pattern, however, the decrease in novice was not statistically significant. No significant changes in urinary catecholamine in any group occurred over the weekend.

The authors concluded that the most significant observations were the highly significant decreases in plasma and urinary free cortisol levels during the TM technique, the effect being cumulative with increased meditation experience. It would appear from this research that the primary effect of the state produced by the TM technique seems to be on the adrenal cortex rather than the adrenal medulla. However, as we shall see in the next chapter, other long term studies have reported distinct changes in adrenalin and noradrenalin levels.

Bujatti and Riederer (1976) analysed urinary metabolites of several major neurotransmitters: serotonin (primary metabolite 5-hydroxyindole-3-acetic acid, 5-HIAA), norepinephrine (primary metabolite vanillic-mandelic acid, VMA) and dopamine (primary metabolite homovanillic acid, HVA). Samples were taken over a four-and-half hour period, from mid-afternoon to early evening, two hours before and two hours after a 30-minute meditation. Controls practicing no form of relaxation were sampled over the same time interval. The study controlled for circadian rhythm effect and diet. The urine concentrations of all metabolites measured in the controls did not change significantly during the experimental period. Meditator samples showed a significant increase in 5-HIAA after meditation, with 5-HIAA concentrations at significantly higher levels than controls throughout the experiment. Resting concentrations of urinary VMA were considerably lower in meditators than controls, but no significant change was found during meditation.

Changes in Neurotransmitter Profile

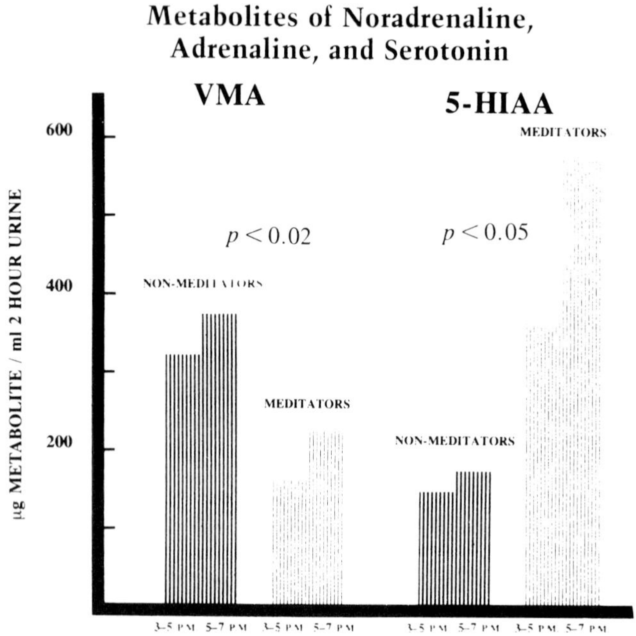

Lower levels of the adrenaline and noradrenaline metabolite, vanillomandelic acid (VMA) and higher levels of the serotonin metabolite, 5-hydroxyindole-acetic acid (5-HIAA) were found in Citizens of the Age of Enlightenment, both before meditation and during the meditation and post-meditation periods, than in non-meditating controls at the same time of day. Lower VMA excretion is consistent with reduced arousal levels, while high 5-HIAA levels are consistent with reports of decreased depression and increased feelings of satisfaction, happiness, and bliss

BUJATTI, M., and RIEDERER, P. Serotonin, noradrenaline, dopamine metabolites in Transcendental Meditation-technique. *Journal of Neural Transmission* 39: 257–267, 1976.

Bujatti and Riederer suggest that the increase of 5-HIAA in TM-subjects indicates an increase in serotonin output by the enterochromaffin (EC) system in the gastronintestinal mucosa, a main site of production and storage of serotonin in mammals. They suggest that the EC system, via serotonin secretion acts as the mediator of a "rest and fulfillment" parasympathetic response produce by the TM technique.

Kochabhakdi and Chentanez (1980) have found an increase in 5-HIAA and also in the amino acid tryptophan, the precursor of serotonin, during the TM technique. They also found that plasma levels of serotonin did not change, thus suggesting that there was an increased serotonin turnover during the TM technique. They also found an increase in white blood cells during and after the TM technique while hematocrit did not change.

In an interesting study by McCuaig (1974), salivary-electrolytes, protein, and pH were measured before, during, and after TM meditation. During meditation there was a general increase in salivary minerals, especially sodium (70%), magnesium (42%), and calcium (36%). Ten minutes after meditation these returned to pre meditation levels. Total protein content also increased during meditation by 60%. Salivary pH decreased during meditation (0.4 pH unit) and increased slightly (0.1 pH unit) one minute afterwards. The decreased salivary pH was thought to reflect the mild metabolic acidosis found by Wallace and coworkers (1971). McCuaig felt that these changes reflected alterations in specific processes involving these substances and could not be completely explained by an overall change in water concentration. A more careful study of salivary proteins would be very interesting since important growth factors are known to be present in the salivary glands.

AUTONOMIC EFFECTS

A number of studies have attempted to measure various parameters of the autonomic nervous system, such as skin resistance and heart rate as they are affected by the practice of the TM technique. In the early studies, a very slight decrease in heart rate was noted along with a marked increase in resting skin resistance. Orme-Johnson (1973) also reported a significant increase in resting skin resistance in 12 out of 13 subjects, while nonmeditating control subjects, during rest with eyes closed, showed nonsignificant changes.

Several later studies have attempted to compare the effects of the TM technique with those of relaxation techniques. In general, the results of these studies show that all these techniques, by causing some degree of relaxation, can effect autonomic measures. We will not attempt to describe the details of these studies. Even though, in some cases, they include a very careful experimental design, they unfortunately utilize only autonomic measures (i.e., dependent variables) which by themselves are not enough to distinguish between different states of rest and relaxation. Also, a number of these studies utilize an insufficient number of subjects to adequately distinguish statistically between different procedures (Orme-Johnson and Dillbeck, 1986).

Any conclusions drawn from these studies are inherently limited due to the fact that these autonomic measures change with almost any type of relaxation. Perhaps the most interesting and valuable use of autonomic measures is in those studies examining the effects of the TM technique on people in challenge situations outside of meditation. These will be discussed in several later chapters.

EEG

EEG is one of the measures used most frequently to study techniques of meditation.

The first EEG studies on the TM technique showed increased intensity (mean square amplitude) of slow alpha waves in the central and frontal regions of the brain, interspersed with occasional high voltage trains of theta in the frontal channels (Wallace, 1970a; Wallace et al., 1971). The intensity of beta and delta waves either decreased or remained constant during meditation. In subjects who reported feeling tired and drowsy at the beginning of meditation a flattening of the alpha activity along with low voltage, mixed frequency waves were observed. As these subjects continued meditation, this pattern was replaced by regular alpha activity. The five subjects measured showed no change in electroculograms.

Banquet (1973) examined 12 TM meditators and compared their EEGs to those of 12 matched non-meditating controls about to learn the TM technique. Banquet used the pre and post meditation control paradigm and also supplied each subject with push-buttons to signal five internal events: body sensation, involuntary movement, visual imagery, deep meditation, and transcendence (the state of pure consciousness). He reported finding three distinct stages of meditation, the first stage being characterized by alpha activity, which increased in amplitude and slowed to 1 to 2 Hz initially in the frontal channel. This pattern was found to repeat at the end of meditation with increased alpha abundance. These alpha periods usually were not blocked in response to flash and click stimuli. A second stage was characterized by a dominant theta pattern. Short theta bursts of high voltage (up to 100 microvolts) were seen simultaneously in all channels (or first frontally), usually followed by longer rhythmic theta trains (10 seconds to several minutes) at 60-80 microvolts.

Banquet notes that the theta activity was unlike that of drowsiness (stage I sleep) in that the pattern was blocked by click stimuli but reappeared spontaneously within a few seconds. In drowsiness, click stimuli produce the alpha rhythm of an arousal reaction. The spectral array analysis revealed morphological differences between meditation and drowsiness. Meditation theta rhythms appeared as continuous trains of a specific frequency. Drowsiness, on the other hand, produced a mixture of discontinuous theta frequencies combined with alpha and low delta.

Four of the meditators showed a third state similar to that reported by Das and Gastaut (1957), which was characterized by fast beta spindle bursts (20 and 40 hz) alternated with alpha or theta rhythms. These bursts reached an amplitude of 30-60 microvolts and maintained a constant frequency. There was a tendency for them to become continuous on a background of slower activity. They occurred predominantly in the anterior channels, but were present (sometimes simultaneously) in all channels. This EEG activity was correlated with experiences of "deep meditation" and "transcendence" and was unaffected by click stimuli. Tonic electromyogram activity (electrode placement below chin) disappeared and breathing became very slow and shallow. Behaviourally, however, the subject was able to respond to questions readily and accurately.

Topographic changes in the EEG were also observed. There was a tendency toward synchronization of anterior and posterior channels. Alpha rhythms spread from occipito-parietal regions forward. Theta and beta frequencies generally appeared in frontal channels first and diffused posteriorly. Transient asymmetry could occasionally be seen between right and left hemispheres in shifting from slow to fast frequencies. There were periods of uniformity of wave form frequency and amplitude in all channels simultaneously.

Banquet noted that the EEG stages during meditation tended to occur in cycles. All stages appeared sequentially again if the meditation time were long enough. In advanced TM subjects, alpha and sometimes theta waves persisted in the eyes open period after meditation.

The control subjects showed changes different from those seen in the TM subjects. Control subjects were divided into two groups. The first group did not develop stable alpha rhythms, but showed beta- dominant mixed frequencies throughout. In the second group, alpha activity appear primarily in the posterior region of the brain and tended to alternate with beta activity. Some members of the second group also showed EEG signs of drowsiness and sleep.

In another paper, Banquet and Sailhan (1974a) performed statistical analyses of the EEG during TM meditation and found significant differences between eyes closed rest and the end of meditation.[1] The data also clearly distinguished the TM technique from sleep and drowsiness.

In addition, they noted that meditation generally produced an increased phase synchrony between adjacent parts of the brain. Two subsequent studies replicated these findings. Kras (1974) found a significant increase in alpha wave activity in all parts of the brain in subjects practicing the TM technique, as compared to relaxing controls. Wescott (1973) found similar increased synchrony in EEG activity within each hemisphere of the brain and

---

1. These data were derived from the average of the power spectra, using the following ratios: depth of sleep coefficient, delta (alpha and theta); wakefulness coefficient, alpha/delta; activation coefficient, beta/alpha. Cross-correlations were also computed between monopolar derivations of a transverse montage.

also a more even distribution in EEG alpha activity between the right and left sides of the brain as compared to controls.

Hebert and Lehmann (1977) investigated the frequency of occurrence of the high amplitude (more then 100 microvolts) theta bursts reported during the TM technique. They examined 78 long-term meditators who had been practicing the TM technique from 24-140 months, with a mean of 56.1 months. All were healthy, normal people, socially well-adjusted with no history of recent drug use, brain damage, seizures or fainting. These were compared to 54 control subjects who had not learned the TM technique. EEGs were recorded both during meditation and sitting quietly with eyes closed.

Of the 78 meditators tested, 21 demonstrated intermittent prominent bursts of frontally dominant theta activity. Theta bursts occurred an average of every two minutes, with an average duration of 1.8 seconds and an average maximal amplitude of 135 microvolts, ranging as high as 300 microvolts. The bursts usually were preceded and followed by alpha rhythm. Subjective reports elicited during the theta burst activity indicated pleasant states with frequent references to "peaceful" or "comfortable" and descriptions of "drifting" or "sliding." The subjects maintained intact situational orientation and reported no experiences related to sleep. The 54 non-meditating controls exhibited no theta bursts either during relaxation or sleep onset. The authors hypothesized that theta burst may be the manifestation of an adjustment mechanism, which comes into play during prolonged low arousal states such as that produced by the TM technique.

EEG COHERENCE

Perhaps the most important development in the EEG research on the TM program has come through the application of EEG coherence measurements. EEG coherence is derived from a computer analysis of the EEG signals from two spatially separated areas of the brain. Coherence provides a measure of the correlation between two EEG records for each frequency and attains a high value at a given frequency if the phase relationship between two channels is nearly constant over a specified time interval.

One of the first applications of this measurement was by Drs. Donald Walters and W. Ross Adey and co-workers working at the Brain Research Institute at U.C.L.A. In a series of articles Walter and co-workers utilized coherence to distinguish between various functional states. For example, they noted a decrease in alpha and theta coherence during sleep. They further speculated on the location of key centers of the brain possibly responsible for the generation and synchronization of EEG signals. One of the most interesting findings of their studies was a marked increase in theta EEG coherence in Astronaut Frank Borman is Gemini Flight, during a 40-minute period in the second half of the first orbit, when he was in very relaxed yet awake state (Walter, et al. 1966, 1967a, b).

Coherence measures have also been used in a number of other studies (Gevins, et al. 1980), for example, by John in his Neurometric testing program, which includes a variety of computerized EEG and evoked potential measures and which has been utilized to successfully distinguish certain child learning disabilities (John, et al., 1970; John, 1977).

In order to examine EEG coherence during the TM technique, Levine (1975) developed a computerized technique known as the coherence spectral array (cospar). Levine used digital filtering techniques designed to register coherence peaks in the array only if they exceeded 0.95 (1.0 being perfect coherence) and if the coherence relationship existed for more than 10 seconds. This procedure biased the cospar in favor of showing relationships of only high and long-term coherence and thereby increased the likelihood that a given coherence peak represented true long range order in the EEG.

Since TM-specific changes had been reported mainly in the frontal and central regions of the brain, cospars were computed only for these areas (electrode placement at F3, F4, and C3, C4) on the 25 subjects studied. The most common effect was an increase in the height and/or incidence of coherence peaks in the alpha band with the beginning of the TM technique but without a marked decrease at the end of that period. The second most frequent effect during meditation was the spreading of coherence peaks to other, generally lower, frequencies. The third was an abrupt onset of strong coherence with the start of the TM technique and an abrupt decrease at the end.

Intrahemispheric TM-specific effects appeared more frequently and dominantly in the frontal channels. One of the most interesting cospars observed was that of a long-term meditator of 15 years, in which continuously strong coherence was seen concurrently in delta, theta, alpha, and beta bands both during and outside of meditation. Levine points out that alpha coherence is not unique to the TM technique, although the patterns of change observed, particularly in the spread of coherence to other frequencies, appear to be characteristic of the TM state. This was not observed in control subjects during either extended periods of eyes-closed relaxation or in "mock" meditations involving repetitive backwards counting in a nontaxing fashion. In the "mock" meditation any strong alpha coherence which may have been present upon closing the eyes initially tended to be reduced.

Levine found that drowsiness and sleep were characterized by a loss of any consistent, strong interhemispheric coherence which may have been present in the alpha band, and a gradual decline in total coherence.

Coherence of the Brain during the Practice of the Sidhis

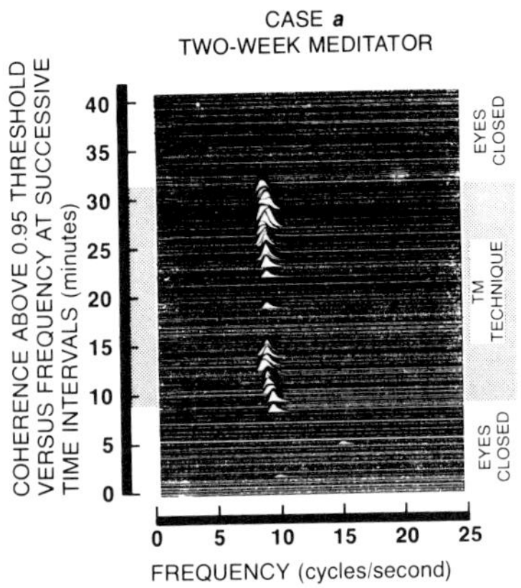

COSPAR: New meditator (two weeks' practice of the TM technique). Note that the strong coherence in the alpha band (near 10 Hz) is specific to the period of the TM technique.

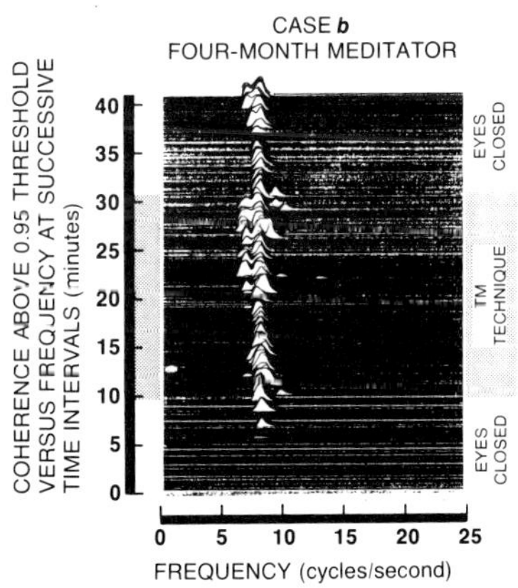

Varieties of coherence increases specific to the TM technique (four months' practice of the TM technique).

BEL, U. EEG coherence during the Transcendental Meditation technique. Psychophysiology Laboratory, Centre for the Study of Higher States of Consciousness, Maharishi European Research University, Switzerland, 1975.

LEVINE, P. H.; HEBERT, J. R.; HAYNES, C. T.; and STRO-

Increased Coherence

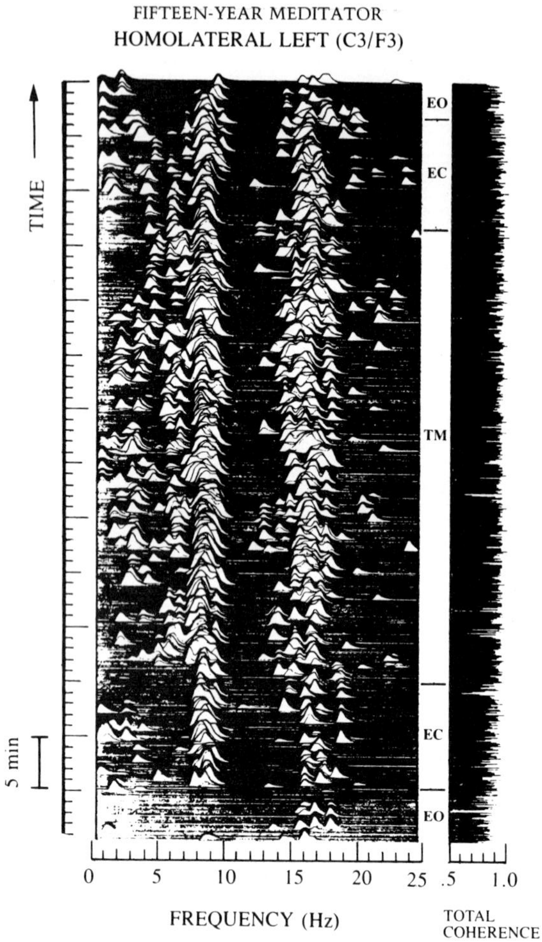

COSPAR of a long-term meditator. In the most experienced subject studied, the highest levels of coherence were found. Note how during the period of the Transcendental Meditation technique the coherence peaks extend over a major portion of the 0–25 Hz band. The strong beta coherence—possibly harmonically related to the alpha/theta activity—is particularly unusual. Strong theta coherence near 6 Hz begins abruptly with the start of the period of the technique (EO = Eyes open, EC = Eyes closed)

In REM sleep there was significant coherence in the delta band along with an increase in total coherence. Also, sleep spindles were sufficiently coherent to be picked up by the cospar. These findings of changes in coherence during certain stages of sleep are consistent with the earlier sleep studies on nonmeditators (Dumermuth et al., 1972). They further help to clearly distinguish the physiological changes seen during the TM technique from those seen in sleep or drowsiness (Levine et al., 1975).

Farrow and Hebert (1982), as part of their extensive analysis of periods of respiratory suspension during the TM program, studied detailed changes in skin resistance, heart rate, and EEG power and coherence in one advanced subject. Again, periods of respiratory suspension were highly correlated with the subjective experience of pure consciousness. While the periods of respiratory suspension were abrupt, discrete, and relatively uniform, changes in other parameters associated with the experience of pure consciousness were graded and more variable. Basal skin resistance increased before and during these periods and often dropped abruptly at the end of the periods. The mean heart rate decreased during the episodes and then increased after the periods.

Inspection of the EEG showed bursts of beta activity immediately after the end of the respiratory suspension periods. Delta band power was usually low and stable before and during these episodes. Averaged signal tracing showed that theta band power peaked sharply at the onset of breath suspension periods and decreased sharply at the end. Changes in alpha band power were relatively large but inconsistent, showing no obvious relationship to the subjective experience.

EEG coherence in the alpha and beta bands was high before and during the first half of the respiratory suspension period, decreased gradually during the second half, and then decreased abruptly at the end of the period.

Coherence in the delta band was much more variable, but also dropped at the end of the respiration suspension period. The mean coherence changes in the theta, alpha, and beta bands were statistically significant. Widespread changes in coherence over many EEG leads were seen. Of the ten EEG coherence derivations studied (F3-F4, C3-C4, F3-C4, F4-C4, P2-O2, T3-C3, T3-F3, F4-O2, F4-C3, F3-C3), theta band coherence exhibited significant change in eight derivations, alpha band coherence in six, beta band coherence in three, and delta band coherence in only one.

In a more recent study (Badawi et al., 1984), the preliminary results of Farrow and Hebert were replicated in a much larger group involving more than 100 subjects and including several additional meditating and nonmeditating control groups. EEG alpha coherence, especially in the frontal areas of the brain, was found to increase during periods of respiratory suspension. The first control group, consisting of nonmeditators showed no periods of respiratory suspension during relaxation with eyes closed. In the second control group, TM subjects were asked to voluntarily hold their breath while EEG coherence was measured. There was no significant change in EEG coherence during these periods. One of the conclusions of this study was that EEG coherence was more sensitive to periods of pure consciousness than were EEG power measures and that further studies should utilize markers such as periods of respiratory suspension in order to more precisely study the physiological characteristics of the pure consciousness state.

From these EEG studies as well as the other physiological studies on breath suspension during the TM technique, it appears there are distinct physiological correlates associated with the experience of pure consciousness. These physiological correlates also clearly indicate that a variety of other possible states may occur during the TM technique. For example, a state of deep relaxation has been reported which is characterized by increased frontal alpha wave activity, the appearance of high voltage theta spindles and increases in basal skin resistance. Also reported particularly in unusual experimental conditions, are occasional periods characteristic of various EEG sleep stages.

The most definitive studies by far are those which have attempted to carefully distinguish the state of pure consciousness from other states. These studies have noted specific periods of low metabolic rate, respiratory suspension, and high intra and interhemispheric EEG coherence in alpha and theta frequencies, especially in the frontal and central areas of the brain, which are highly correlated with the subjective experience of pure consciousness.

By using more careful subject selection, coupled with subjective reports, simultaneous electrophysiological measurements, and/or more in-depth physiological measurements, a more precise analysis of the state of pure consciousness has been obtained.

REFINEMENT OF METHODOLOGIES TO STUDY THE LEAST EXCITED STATE OF CONSCIOUSNESS

In the most thorough review article on all the physiological studies on the TM technique, Orme-Johnson and Dillbeck (1986) used meta-analysis to compare the state produced during the TM with ordinary rest or relaxation. Their analysis clearly distinguished the state produced during TM from rest and indicated several methodological problems in previous review articles.

The first problem has been a tendency to aggregate data from different meditation techniques that are procedurally quite different and have different effects.

Second, previous reviews do not always fully utilize

the information present in the studies of meditation that were cited. Two possible approaches can be taken, narrative and quantitative, each giving quite different conclusions. In the narrative approach, some studies are excluded on design considerations and the remaining results verbally integrated. This can lead, unfortunately, to a great degree of variability due to a number of factors such as different criteria of what is a "good" study, differences in interpretation of results, and use of crude criterion measures such as presence or absence of statistical significance while ignoring such considerations as statistical power.

The importance of statistical power and sample sizes is a critical issue given the sample sizes of many of the studies on meditation. As Orme-Johnson and Dillbeck point out, for a moderate effect size to be detected, there should be a minimum of 15 subjects per group for between-subject comparisons and 10 subjects per group for within-subject comparisons.

In their review, Orme-Johnson and Dillbeck adopted a quantitative approach in which the statistical measure known as effect size was calculated. In order to do so they first completed a series of computer searches, locating all studies listed in the following data bases through 1985 with the keywords "meditation" or "relaxation response": *Psychological Abstracts, Science Citation Index, Social Science Citation Index, Index Medicus, and Sociological Abstracts.* The reference section of each paper listed was also searched for additional citations. Studies that assessed physiological effects of the TM technique were selected. In addition, other papers also listed in previous reviews were included for their measures of change during the eyes-closed rest control condition.

Orme-Johnson and Dillbeck then calculated effect sizes for each of the studies. Because they wished to make use of all the information possible from all studies, effect sizes were calculated separately for the TM technique and rest as the number of standard deviations of change from the pre-meditation or pre-rest mean (using the standard deviation of the pre-period).

Using the statistical technique of meta-analysis on physiological research on the TM technique they demonstrated that the effect size for the TM technique is significantly larger than for ordinary eyes-closed rest for findings of increased basal skin resistance and decreased respiration rate, oxygen consumption, and plasma lactate. No significant difference between eyes-closed rest and TM was found for heart rate or spontaneous skin resistance responses, although TM participants were found to have significantly lower levels of these two variables, as well as lower respiration rate and plasma lactate levels, during the pre-meditation or pre-rest baseline periods.

Orme-Johnson and Dillbeck made two major recommendations for future research. The first was methodological. Studies comparing the TM technique and rest should have an adequate sample size, should be designed with explicit calculation of statistical power, and should also take closer account of the relevant confounding variables (e.g. dynamics and sub-stages of the TM technique, individual differences, recent and long-term life experiences of stress, differences over time in the quality of experience, and length of time practicing the technique.)

The second recommendation was that researchers studying the effects of the TM technique should be familiar with the traditional theoretical framework of the technique. In particular, the three most important concepts for those studying this technique at present are : (1) there is a state of transcendental consciousness that is predicted to have unique physiological correlates, including global physiological integration characteristic of a restfully alert state rather than just reduced somatic arousal; (2) the TM technique is best viewed as a dynamical process with alternating sub-stages of transcendental consciousness and physiological normalization; and (3) the regular practice of the TM technique is predicted to develop better health and adaptive efficiency.

In summary the refinement of methodologies and measurements utilized, as well as the increase in the number of studies, has helped provide a more detailed description of the physiological correlates of this least excited state of consciousness, as well as a deeper understanding of basic neurophysiological mechanisms involved. This more complete objective physiological description of pure consciousness the fourth major state of consciousness, is an extremely important step in the integration of modern science and ancient Vedic Science. For the first time the concept of the development of higher states of consciousness has been taken out of the realm of philosophy and mysticism and shown to be a scientific reality. We now have the foundation for objectively understanding the neurophysiological basis of growth toward the state of enlightenment.

Chapter 4

THE DEVELOPMENT OF HIGHER STATES OF CONSCIOUSNESS

If the nervous system is capable of supporting a least excited state of consciousness, is it then also capable of supporting higher or more optimal states of consciousness? As we will see in a later chapter, numerous studies have demonstrated that experience affects the structure and function of the nervous system. The presence or absence of certain types of stimuli during critical stages of animal development has a dramatic effect on the basic anatomical connections of nerve cells in the brain and thus on all types of perceptual and behavioural processes. In the light of these experimental findings, it is appropriate to ask: what is the effect of the presence or absence of the experience of pure consciousness on the functioning of the nervous system?

To answer this question we must consider the empirical evidence concerning the effects of the Transcendental Meditation program not only during meditation, but during activity, and further, the possible underlying neurophysiological mechanism for the development of higher states of consciousness. According to the Vedic tradition, the development of these higher states occurs as a result of the least excited state being spontaneously integrated into activity.

Maharishi, in his revival and systematic presentation of ancient Vedic wisdom in the Science of Creative Intelligence and Vedic Science, has described the process of development of optimal neurophysiological functioning in terms of seven major states of consciousness. The first three are the familiar waking, dreaming, and sleeping states of consciousness. The fourth state is the least excited state, pure consciousness. According to Maharishi, with the regular experience of pure consciousness the nervous system adapts to a new style of functioning. There is a gradual refinement of neurophysiological functions resulting in a new, more expanded state known as cosmic consciousness, in which pure consciousness exists along with the other three states. Thus, even when the individual is completely asleep he experiences a kind of inner alertness that is often referred to by various traditions as "witnessing."

Maharishi then describes how further refinement of neurophysiological functioning results in two even more optimal states, known as refined cosmic consciousness and unity consciousness. The Vedic records give very clear and definitive descriptions of these states. In the absence of a systematic procedure to experience them, however, they have remained obscure and virtually unobtainable. With the introduction of the Maharishi Technology of the Unified Field, a systematic technology of consciousness, it has for the first time become possible both to develop and experience higher states of consciousness on a wide scale in the general population, and to conduct scientific research on the more optimal neurophysiological functioning that supports them. In this chapter we will examine research specifically concerned with physiological and psychological measurements in TM meditators outside of meditation, and attempt to develop a physiological model for the growth of higher states of consciousness. In the following chapter we will continue this examination, focusing only on studies concerned with the TM-Sidhi program.

HOMEOSTASIS—PHYSIOLOGICAL ADAPTABILITY AND STABILITY

From the point of view of a physiologist, the evolution or optimization of biological functioning is most clearly seen in terms of the development of more sophisticated homeostatic mechanisms. The concept of homeostasis, first developed by the eminent French physiologist Claude Bernard and later expanded by the American physiologist W.B. Cannon, refers to the ability of a living organism to maintain internal stability and order by adapting to external change. Virtually every system in the human body utilizes some type of homeostatic feedback mechanism. As a result, many internal physiological conditions such as temperature, acid-base balance and blood sugar levels remain stable even amidst wide fluctuations in environmental conditions and various demands or challenges placed upon the system. Research on the development of higher or more optimal states of consciousness through the TM program has demonstrated an improvement in a number of homeostatic adaptive processes.

Orme-Johnson (1973) conducted one of the most important early studies on the effects of the TM technique outside of meditation, which clearly demonstrates improvements in physiological stability and adaptability. This study showed that TM meditators had fewer spontaneous skin resistance responses (100 ohms or greater) outside of meditation than did non-meditating controls. Orme-Johnson interpreted this finding to indicate a more stable and optimal style of autonomic functioning in the TM subjects as a result of their practice of the TM technique.

Previous studies had shown that individuals with low levels of spontaneous skin resistance responses and heart rate, show less motor impulsivity; that is, they are less likely to make impulsive and inappropriate movements during a simple motor test like reaction time. Also these

Spontaneous Skin Resistance Responses

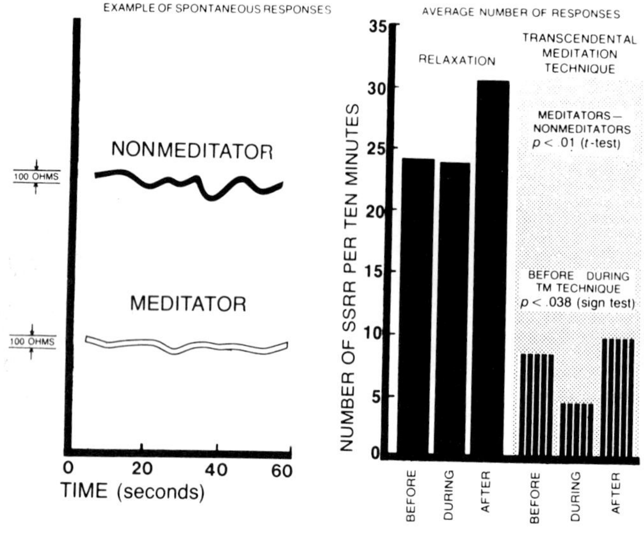

Finding: Subjects practicing the Transcendental Meditation technique were found to have fewer spontaneous skin resistance responses (SSRR) than non-meditating control subjects ($p<.01$), indicating greater stability in the autonomic nervous system.

Reference: David W. Orme-Johnson, "Autonomic Stability and Transcendental Meditation," *Psychosomatic Medicine* 35, no. 4 (U.S.A.: 1973): 341–349.

Habituation of Skin Resistance Responses

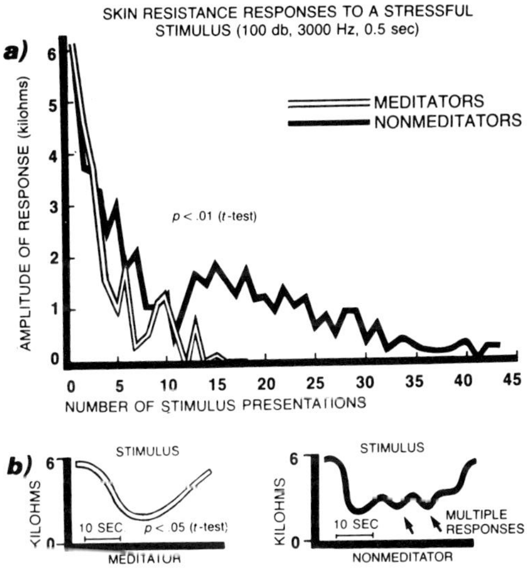

Finding: In this study induced changes in skin resistance in response to a stressful stimulus were measured. Both the wave form of the individual response and the degree of habituation (reduced response) to repeated stimuli were recorded. Subjects practicing the Transcendental Meditation technique were found to habituate more rapidly to a series of auditory stresses (loud noises) than non-meditators (figure *a*). In addition, the wave form of the response to the first stress was significantly smoother and more stable in the meditators (figure *b*).

Reference: David W. Orme-Johnson, "Autonomic Stability and Transcendental Meditation," *Psychosomatic Medicine* 35, no. 4 (U.S.A.: 1973): 341–349.

individuals were better able to withstand stress.

Orme-Johnson's study also included measurements of how quickly the skin resistance response adapts or habituates to stressful stimuli in TM subjects as compared to controls. Habituation is defined as the systematic decrease in response amplitude to the repeated presentation of the same stimulus. Previous research indicates that an individual who habituates rapidly on skin resistance response is outgoing and has a stable nervous system. Orme-Johnson measured evoked skin resistance responses in meditators and non-meditators who were presented with noxious tones (100 db, 0.5 sec., 3000 Hz) at regular intervals. The TM meditators habituated significantly faster than the non-meditators (mean of 11 trials, as compared with 26.1 trials). Also, meditators produced fewer multiple fluctuations in skin resistance during the recovery cycle, thus suggesting a more stable reaction to stress.

Goleman and Schwartz (1976), using continuous measures of heart rate and phasic skin conductance, as well as self-report in personality scales, also investigated the response pattern to stressful stimuli of TM meditators as compared to a relaxation control group. The respective groups either meditated or rested after they viewed a film of 50 workshop accidents. The meditators' heart rates increased more than the non-meditators' in anticipation of the accidents, but recovered more quickly than the non-meditators' after the accident scenes. Phasic skin conductance responses also increased more in meditators with anticipation of the accidents, but again habituated more quickly following the accident scenes. Meditators also reported experiencing less subjective anxiety during the experiment.

Several other studies have repeated Orme-Johnson's findings of both increased autonomic stability, as measured by a lower number of spontaneous skin resistance responses in TM participants as compared to controls (Wilcox, 1973; Berker, 1974; Smith, 1974), and faster habituation of autonomic responses such as evoked skin resistance response and heart rate (Graham et al., 1973).

One very extensive study by Daniel (1976) involving 170 subjects studied the effects of a stressful stimulus (the sound of teeth grinding or the sound of a finger nail being drawn down a blackboard) on galvanic skin response (GSR) in TM subjects outside of meditation as well as in a variety of carefully chosen control groups. Six main groups were involved: group 1 had been regularly practicing the TM technique for more than three years, group 2 had been practicing the TM technique from six weeks to two-and-one-half years; group 3 had been practicing progressive relaxation for at least six weeks; group 4 did not add anything to their daily routine; group 5 were practicing Hatha Yoga (any posture they desired); and group 6 subjects were practicing a simulated meditation using nonsense syllables (thought up by the experimenter). Each of these groups was then divided into three subgroups of approximately 10 subjects: an experimental group, who were told before the experiment what noises they would hear; a control group, who were not told what noises they would hear; and a control group, who did not hear the noise but were asked to imagine it.

The results indicated that the long-term participants in the TM program showed very little response to the disturbing noises and seemed to have almost total control of GSR. Further, the TM subjects who had been meditating the least amount of time showed significantly fewer responses during the test period than the control groups. The study also showed that the TM meditators as compared to non-meditators recovered more quickly from stressful stimuli, thus reconfirming Orme-Johnson's original results.

Another study, by McDonagh and Egen (1973),

examined the effects of the TM technique on temperature homeostatic mechanisms. Skin temperature was measured in TM meditators and non-meditating controls before, during, and after a period of running in place for five minutes. In both groups the skin temperature was initially stable and dropped during running in response to the exercise. After running, the skin temperature of the TM group showed a significantly more rapid return to the initial baseline levels than that of the non-meditating controls. The results again suggest that the practice of the TM technique results in a faster and improved adaptive response to stressful stimuli.

In a study reported at the International Conference on Research on Higher States of Consciousness at Mahidol University in Bangkok, Thailand by Yee and Dissanayake (1980), a two-hour glucose tolerance test was administered to subjects before they began the practice of the TM technique and then again after two months of regular daily practice. The glucose tolerance test involves the intravenous injection of a solution of glucose in water. Blood samples are collected immediately after the injection and at various intervals in order to evaluate the ability of the pancreas (via insulin secretion) to return the abnormally high levels of glucose in the blood to their normal value. The response declines with age, and blood glucose levels are higher and take longer to return to the normal value. After the subjects had learned the TM technique, their blood glucose levels were not as high immediately after the injection as in the premeditation control measurements, and returned to normal more quickly. Thus the TM technique appears to significantly improve the homeostatic adaptive mechanism responsible for maintaining glucose stability.

Several other studies have looked at the effects of the TM technique on the resting levels of certain metabolic and physiological parameters. Routt (1973) measured heart rate, respiratory rate, skin resistance, heart rate variability, and finger-pulse volume in TM subjects outside of meditation, as compared with an equal number of non-meditating controls. He found significantly lower levels in heart and respiratory rate and a significantly higher level in skin resistance in the TM subjects as compared to the control subjects. No significant difference was found in the heart rate variability or finger pulse volume.

Lin and Chandra (1980) found that after one month of the regular practice of the TM technique, Chinese subjects showed reductions of metabolic rate, respiratory evaporative heat loss, and whole body conductance. Cold tolerance and auditory acuity each markedly increased. There was no change in visual acuity or body temperature.

In our Aging Laboratory at Maharishi International University, we measured systolic blood pressure in 112 subjects practicing the TM and TM-Sidhi program (Wallace et al., 1983a). The subjects were between the ages of 35 and 64. The TM and TM-Sidhi subjects had significantly lower systolic blood pressures than norms for the general population (matched for sex, race, and general educational background). This difference was independent of diet and exercise patterns but related to length of time meditating. Long-term participants practicing of the TM program (practicing more than five years) showed a significantly lower level as compared to short-term participants (less than five years), covarying for age.

The overall conclusion of all of these different studies taken together is that the TM technique appears to improve both the stability and adaptability of homeostatic adaptive mechanisms. How can we account for this improvement? One possible explanation is that as the nervous system becomes freer from stress, its stability and sensitivity increase. If we improve the electronics of a radio or TV and remove the internal static, the reception is much sharper and clearer. If we refine and integrate the functioning of the nervous system during the TM technique, the "internal noise" of the system is reduced, and adaptability and flexibility in activity will inevitably increase; one will naturally be more capable of responding to the demands of the environment and quickly returning to an optimal homeostatic resting point. Greater external adaptability promotes greater internal stability, and vice versa. This is seen most clearly in TM meditators' increased ability to maintain a more stable and restful physiological state simultaneously with their improved performance on various perceptual-motor tasks. By operating in a smoother and more orderly style of physiological functioning, the internal disorder or entropy of the system is reduced, enhancing perceptual and motor functioning.

PERCEPTUAL AND MOTOR PERFORMANCE CHANGES

A number of studies have shown significant improvements in various aspects of perceptual and motor performance with the practice of the TM technique. Shaw and Kolb (1971) found that reaction time, which remained static or increased after mere relaxation, decreased after the practice of the TM technique. Appelle and Oswald (1974) reported that subjects practicing the TM technique displayed faster reactions than matched control groups. Holt and coworkers (1978) also noted that the TM technique, improved responses on visual choice reaction time. Rimol (1974), using the Swedish Labyrinth game, found similarly improved performance on a perceptual-motor task in subjects practicing the TM technique. Blasdell (1977) found subjects made fewer errors and were faster on the mirror star-tracing test as a result of practicing the TM technique when compared to non-meditating controls. Martinetti (1976) reported that TM meditators showed a significant reduction of perceptual illusion and Dillbeck (1982), in a very well controlled and extensive study, showed longitudinal improvements in visual perception and verbal problem solving. In several different studies (Pirot, 1973; Schwartz, 1979), short and long term improvements have been reported on hearing ability and discrimination.

A study by Schwartz (1979) is of particular interest since it includes measures of auditory thresholds and reaction time as well as several other parameters. Schwartz's study involved thirty volunteer subjects ranging in age from 18 to 33. Fifteen subjects who had practiced the Transcendental Meditation technique for a minimum of five years were tested three separate times in a six-week period. Fifteen non-meditating controls were tested twice and then were instructed in the TM technique. Three weeks and six weeks following instruction in TM, the subjects were retested. Schwartz assessed simple finger reaction time to continuous pure tones at 40 dB, 55 dB, 70 dB, 85 dB and 100 dB. In each test session, each subject was given three blocks of reaction time trials. Each block consisted of five trials at each of the five stimulation levels. Auditory thresholds were assessed and time estimation was examined four times in each test session.

With the exception of time estimation, all variables showed significant change from session one to session two in the long-term TM group as compared to nonmeditating controls. Analysis of reaction time data for sessions one and two indicated that, of the two groups, the long-term TM meditators had significantly faster reaction times at every level of stimulation. The long-term TM group also showed significantly lower auditory threshold levels than did the controls during sessions one and two. After session two the control group learned the TM program, which they practiced twice a day for the remainder of the study. Analysis of variance showed that, after starting TM, the group improved significantly on both reaction time and auditory threshold.

Reddy and coworkers (1974) have also found marked

improvements in perceptual-motor performance in athletes as a result of practicing the TM technique. Thirty athletes were randomly assigned to an experimental group starting the TM program and a control group. The study found significant improvement after six weeks in several measures of athletic ability and agility, as well as reaction time, coordination, and respiratory and circulatory efficiency. Anecdotal evidence of improvement in perceptual-motor ability has also been given by a number of professional athletes, such as Willie Stargell and Joe Namath, who practice the TM technique.

In an extensive study by Daniels (1976), performance in dichotic listening was measured in fifty subjects. The subjects were divided into five groups: (1) subjects practicing the TM technique; (2) subjects practicing the TM technique but told beforehand that TM subjects perform poorly at this particular task (demotivation); (3) subjects practicing a simulated meditation with the use of a nonsense syllable; (4) subjects practicing program restoration; and (5) subjects who relaxed in their own individual way. Subjects were required to pick out target words embedded randomly in a piece of text. Into each ear separate texts were read at a rate of 180 words per minute. Subjects were tested before and after practicing their specific technique, and scores were determined by the percentage difference in the number of correct hits (positive) on the target word and the number of responses made to the wrong words (false positives). Both groups practicing the TM technique showed significantly better performance on this task than the other three groups. Further, the analysis of false positives indicated that the TM groups performed at a more efficient level before the period of the TM technique than all the other groups before they practiced their own techniques. That the second TM group performed significantly better than the other non-TM groups showed that the improvement in the ability to process information occurred independently of a high motivational factor.

Pelletier (1974) demonstrated, in a longitudinal study using three measures of field independence, that individuals practicing the TM technique show cumulative positive effects on perceptual ability. Field independence reflects the development of a stable internal frame of reference. Some researchers feel that the improvement in field independence is a concrete example of Maharishi's description of the development of higher states of consciousness, in which the state of unbounded pure consciousness is maintained as a stable internal reference even in the midst of dynamic activity.

One further area of sensory and motor physiology is the study of reflex activity. Warshal (1980) studied the short and long-term effects of the TM technique on normal and Jendrassik patellar reflex times. Previous studies have shown that the Jendrassik maneuver (which involves clenching the fists) facilitates the normal reflex and results in shorter reflex times. The normal patellar reflex time was determined by measuring the time between the striking of the patellar tendon below the knee cap with a hammer and the movement of the foot, as measured by the release of a microswitch embedded in a heel plate. In this particular study a fractionation procedure was used in which the total reflex time was divided into two components: neural (latency) and muscular (motor). The neural component measures the time from the stimulus onset to the appearance of action potentials at the knee exterior muscle (restus femoris). The motor component measures the time from the onset of the action potentials to the knee extension phase of the reflex.

Fourteen male subjects of a mean age of 21 years, initially non-meditators, participated in the study. The design used to measure long-term effects involved eight test sessions, the first three occurring on consecutive days before the group learned the TM technique and the next five occurring after starting the TM technique at one-week intervals. A high degree of test-retest reliability was observed during the baseline sessions.

The design used to study short-term effects involved measuring reflex times before and after 20 minutes of rest in one of the three baseline periods and before and after 20 minutes of meditation in each of the five sessions after learning the TM technique. The short-term measurements showed no significant difference in reaction time before and after 20 minutes of the TM technique as compared to rest; the long-term measurements, however, showed a significant effect over the duration of the experiment. The normal reflex time for the first three sessions (prior to instruction in the TM technique), as well as its latency and motor components, improved significantly in the following five sessions. There was a reduction of 14.14 msec. in normal reflex time—3.19 msec. in the latency component and 10.95 msec. in the motor component. The findings for the Jendrassik reflex time were similar, with a reduction of 8.72 msec. (comparing the baseline sessions with the last session after starting the TM technique).

Warshal offered the following four possible explanations for the faster reaction times observed as a result of the practice of the TM technique: (1) increased speed of neural conductance; (2) increased neurotransmitter release with decreased synaptic time; (3) greater sensitivity of the muscle spindle; and (4) supraspinal excitation of the spinal motor pool. Several other studies on reflex measures which consider the possible underlying neurophysiological mechanisms produced by the TM technique will be discussed in the following chapter.

EEG TRAIT EFFECTS— BECOMING FULLY AWAKE

Many researchers distinguish EEG state effects—that is, short-term physiological changes specific to a particular state (such as those occurring during the practice of the TM technique)—from EEG trait effects—changes which seem to be characteristic of long-term development in the physiology. A number of EEG and psychological studies suggest that the regular practice of the TM technique results in particular trait effects, which may best be characterized as increased wakefulness and higher mental functioning.

Williams and West (1975) examined EEG responses to intermittent photic stimulation, comparing subjects experienced in the TM technique to a non-meditating control group. Measurements were taken when subjects were not meditating, but were sitting with eyes closed, resting. Alpha blocking and induction were scored from the EEG paper record. The meditators showed a significantly smaller decrement in alpha activity and alpha blocking response than did the controls. Meditators' alpha induction also occurred more frequently and earlier than that of the controls.

Wandhofer and Plattig (1973), Kobal and coworkers (1975), and Wandhofer and coworkers (1976) reported a significant shortening of latencies for auditory-evoked potentials in subjects before the TM technique as compared to lightly sleeping controls. Amplitudes for the N1 peak were significantly smaller in the TM group but larger for peaks P2 and N2. The DC-shift was found to be larger and the on-effect significantly smaller for meditators as compared to dozing controls. The authors note that the meditating subjects could maintain a more "uniform state of vigilance" that optimized the evoked potential recordings.

Banquet and Lesevre (1980) also found that TM participants outside of meditation showed larger amplitudes and shorter latencies of the visual evoked cortical response at N120 and P200, compared with controls. In addition, TM participants showed shorter reaction times and made fewer mistakes on a visual choice reaction test.

Bennet and Trinder (1977) studied EEG activity in TM meditators and found that during meditation the distribution of alpha activity was symmetrical between the hemispheres of the brain as compared to controls, while during tasks the alpha distribution became more asymmetrical in TM subjects compared with controls. There is normally a different hemispheric distribution of alpha according to the type of task given—e.g., in analytic tasks the left side of the brain is more focused, as show by beta activity, while the right side is more relaxed, as shown by alpha activity. On the other hand, during spatial tasks, alpha increases in the left hemisphere. In control subjects, hemispheric distribution of the amount of alpha was not as clearly defined according to the specific tasks as it was in the TM subjects. The increase in hemispheric lateralization (more alpha in either the right or left hemisphere) during tasks specific to each hemisphere of the brain suggests that the state effect of increased EEG coherence during the TM technique leads to a trait effect of increased ability to differentiate and focus outside of TM.

Dillbeck and Bronson (1981) measured EEG alpha coherence and slow alpha power from frontal and occipital derivations during relaxation or TM in 15 subjects. Subjects were tested before and after a two-week baseline period in which half of them practiced twice-daily relaxation and half did not change their schedule. All subjects were then instructed in the TM technique and were retested after two weeks of twice-daily TM practice. The results of the study showed that during the baseline pre-TM technique period there were no group differences and that after the two-week TM technique period subjects showed a significant increase in frontal alpha coherence, indicating a short-term longitudinal effect of the TM technique. The authors did not find longitudinal changes in alpha power. They therefore concluded that alpha coherence is clearly a more sensitive discriminator of the effects of the TM technique than alpha power and should be utilized when attempting to distinguish the TM technique from general relaxation.

As mentioned earlier, Maharishi explains that in the development of higher states of consciousness, the state of pure consciousness begins to coexist and be maintained along with other states of consciousness, even during sleep. Several researchers have studied the sleep EEG pattern of advanced TM meditators to test how the growth of higher states of consciousness might be reflected in these measurements. In one preliminary study of two advanced meditators, Banquet and Sailhan (1974b) reported finding unusual brain activity during sleep. All cycles and phases of normal sleep were present, yet at the same time, Banquet says, the "higher levels of consciousness were unimpaired." The EEG patterns were the same as in normal sleep, with the addition of a rhythmical beta pattern at a constant frequency (20 Hz) present through the different stages. Subjects "were able to perceive and perform a motor act and were conscious of their dreams."

In a more extensive study Banquet and coworkers (1977) measured EEG over ten nights of sleep in five subjects who practiced the TM technique and five control subjects, using standard derivation O1-O2, C3-C4, F3-F4, surface EMG and two EOG electrodes. Analysis involved traditional sleep scoring as well as use of coherence spectral array. Average sleep duration for TM subjects was 64 percent of that of controls and number of sleep cycles was 1.4 versus 2.8 for controls. Sleep stages shortened both in absolute length and percent of total; for example, stage IV time of TM subjects was 16 percent of that of controls; stage I was the only stage significantly increased for TM subjects. Another study by Miskiman (1972a) reported that TM participants who were deprived of sleep needed a shorter recovery period than controls.

Perhaps the most striking changes in EEG trait characteristics, as we shall see in the next chapter, appear to develop as the result of the TM-Sidhi program. These studies on the TM-Sidhi program further demonstrate that longitudinal changes in EEG coherence are correlated with improved psychological performance and seem to be indicative of a more awake and comprehensive style of neurophysiological functioning.

BIOCHEMICAL MEASUREMENTS

Our understanding of the biochemical basis of mental states has greatly expanded in the past few decades. As a result, most current theories of mental health are based upon the balance of key neurotransmitters in the brain. Studies on the effect of the TM program on biochemical measurements, especially on levels of neurotransmitters and their metabolites, are highly interesting and have laid the foundation for a more complete understanding of the biochemistry of higher states of consciousness.

Lang and coworkers (1979) studied biochemical as well as autonomic measures in TM subjects. They found urinary catecholamine levels in advanced meditators (teachers of the TM technique who had been meditating more than two years) to be higher than in most beginning meditators (practicing two years or less). After the practice of the TM technique heart rate was lower and plasma noradrenalin was higher in advanced as compared to beginning meditators. In an exercise session following meditation, noradrenalin (but not adrenalin) levels increased in beginning practitioners while catecholamines in fact decreased in advanced meditators.

Other researchers, including myself, have hypothesized a decrease in sympathetic tone as a result of the TM technique. Contrary to these models, Lang's group suggests that their results indicate an increase in both sympathetic and parasympathetic tone in advanced practitioners of the TM technique; the higher plasma noradrenaline levels indicate increased sympathetic tone, yet the lower heart rate indicates an overriding inhibiting effect on the heart by increased parasympathetic tone. This model suggests the ancient Vedic description of the development of higher states of consciousness as an extension of the state of restful alertness experienced during meditation. The increased parasympathetic tone corresponds to increased restfulness, and the increased sympathetic tone corresponds to increased alertness in the nervous system. I personally find the results and model of Lang and coworkers extremely interesting, and much of my own current research is concerned with extending their results and developing a more dynamic model of the long-term effects of the TM technique.

Subrahmanyam and Porkodi (1980) conducted a very extensive six-month longitudinal study on medical patients before and after starting the TM program, and on normal subjects who were short- and long-term TM practitioners. The medical patients had a variety of disorders—mild aggression, mental retardation, and epilepsy. Both psychological testing and clinical evaluation of the patients showed a number of beneficial effects of the TM technique. For example, there were improvements in the social behavior of the mildly aggressive patients, in the IQ of the retarded patients (a mean increase of about 15 points, from 40 to 55), and in the EEG and clinical condition of the epileptics.

Extensive biochemical measurements were taken in all subjects. After six months, the urinary levels of cortisol and metabolites, several catecholamine metabolites (VMA and MHPG), and the serotonin metabolite 5 HIAA decreased in both beginning and long-term TM meditators. In hyperaggressive patients, cortisol, VMA and MHPG decreased from abnormally high levels; behavioural aggression also decreased. The mentally retarded patients showed increases in cortisol, MHPG, HVA, and 5-HIAA from abnormally low levels. Epileptics showed fewer spikes in the EEG and increased 5-HIAA in the cerebral spinal fluid. In patients without major disorders, cortisol, serum cholesterol, and sugar levels decreased. One of the most interesting results of the study was the fact the different groups showed different types of changes. Vedic Science predicts that the effect of the TM program is to normalize mental and physical health. Thus, depending on the initial abnormal condition, different biochemical levels in different people might change in order to restore balance to the individual system.

Walton and coworkers studied extensively the effects of the TM program on levels of the serotonin metabolite 5-HIAA, with particular emphasis on how the circadian rhythm of urinary 5-HIAA may be influenced by TM practice (Walton et al., 1981). Average data from 10 subjects as well as rhythm analyses of longitudinal data from one subject indicate that the acrophase of the circadian rhythm tends to occur at the fixed time of the afternoon practice session of the TM and TM-Sidhi program. When the number of practice sessions was changed from one in the morning and one in the afternoon to two in the morning and two in the afternoon, the 24-hour rhythm was replaced by a 12-hour rhythm, and the mean 5-HIAA excretion rate was reduced.

Walton and coworkers suggest that the twice-daily practice of the TM and TM-Sidhi program acts to synchronize the circadian rhythm of urinary 5-HIAA excretion. This initial research on the effects of the TM program on circadian rhythms indicates an entire new frontier of research. Walton and coworkers, as an extension of their studies on 5-HIAA, have also initiated a fascinating investigation of a new biochemical substance produced during the TM technique and the more advanced TM-Sidhi program, which they refer to as substance M. We shall discuss this research further in the next chapter.

O'Halloran and coworkers (1985) reported a phasic

conditioned response of the pituitary hormone arginine vasopressin (AVP), yielding 2.6 to 7.1 times normal plasma concentration in association with the TM technique in long-term practitioners (six to eight years of regular practice). The researchers noted that such a large phasic response had not previously been reported, and that the elevation was not accompanied by an elevation of plasma osmolarity, suggesting the effect was not involved with the normal role of this hormone in water regulation. They also reported that a separate group of nonmeditating control subjects studied in the same manner in unstylized ordinary eyes closed rest showed normal plasma AVP concentration.

Other measures of both the TM and resting control groups included: galvanic skin resistance (GRS), muscle metabolism, EEG, and the Spielbeger Anxiety Inventory (STAI). There was a significantly larger increase in GSR in experimental subjects during the TM technique as compared to controls during rest. Also, forearm oxygen consumption (mostly due to muscle) declined markedly during the TM technique as compared to a slight decline accompanying rest. Scores on the STAI were significantly lower for the TM group as compared to controls. EEG recording showed that alpha activity predominated during the TM technique and rest, with approximately 10 percent of the time in Stage I sleep. Finally, measurements of another posterior pituitary hormone, oxytocin, were normal and unchanged during the experiment in both TM and rest subjects.

In the discussion of the results, O'Halloran and coworkers note that the AVP values in the TM subjects were unusually high. This elevation was not the result of a known endogenous rhythm because AVP does not exhibit circadian variation during light hours in mammalian species. The elevation of AVP levels in the TM group was transient, indicated by normal levels of AVP found in samples taken at a different time of day in each meditating subject. Since subjects did not exhibit obvious signs of excess water retention, it seems probable that the elevation of AVP in these subjects is associated with their routine meditation practice.

They noted that the high levels of AVP prior to the actual start of meditation practice (-15 min. sample) suggested a process of conditioning. In considering possible mechanisms, they point out that because of reported results of other physiological variables, it is unlikely that the high levels are due to either changes in AVP clearance, plasma osmolarity or drop in blood pressure. Several measures obtained in the study also rule out contribution of psychologic stress: GRS data indicated relaxation during the eyes-closed period for both groups; EEG activity recorded suggested relaxed wakefulness; marked decline of forearm muscle respiration in the meditation group indicated muscular relaxation; and before and after the experiment, STAI sores for the meditation group were significantly lower than those of the rest group.

Since the increased AVP secretion is not associated with either homeostasis or stress, O'Halloran and coworkers propose that it is related to the behavioral effects of meditation practice. Previous report have shown an ability of peripherally-administered AVP (as well as synthetic AVP analogs) to affect learning and memory. Based upon evidence of this ability of peripherally-administered AVP and its analogs to modify human and animal behavior, they speculate that increased peripheral AVP activity may mediate the reported beneficial effects of TM on learning, memory, and psychotherapeutic processes.

PSYCHOLOGICAL MEASUREMENTS

A large number of psychological studies have been conducted on those practicing the TM technique in a variety of different areas, such as: personality, social behavior, rehabilitation, intelligence, learning ability, recall, academic performance, creativity, and self-actualization. While it is well beyond the scope of this book to review the many psychological studies on TM participants, a brief survey is useful.

One of the most common findings in regard to personality measures is a decrease in both state and trait levels of anxiety after subjects start TM (Davidson et al., 1976; Dillbeck, 1977; Ferguson and Gowan, 1976; Goleman and Schwartz, 1976; Hjelle, 1974; Lazar et al., 1972; Penner et al., 1974; Shapiro, 1974; Shecter, 1975; Zuroff and Schwartz, 1978).

Eppley and coworkers (1985), in an excellent comparative study using meta-analysis, have clearly shown that the changes in anxiety seen in TM subjects are significantly different from those seen in control subjects practicing different relaxation techniques. Such findings as decreased neuroticism, depression and aggression and increased self-esteem, moral reasoning and tolerance have also been reported in TM subjects (Ferguson and Gowan, 1976; Shapiro, 1974; Tjoa, 1975; Nidich, 1975; Van den Berg and Mulder, 1976; Shecter, 1975).

Studies have also reported that TM subjects show greater job satisfaction, improved performance, and better relations with coworkers and supervisors in an industrial setting (Frew, 1974; Friend, 1975).

In the area of nonverbal fluid intelligence, Tjoa (1975) showed that nonverbal, analytic, and logical reasoning increased significantly in TM participants over 12 and 16-month periods. In a study by Schecter (1977), high school students were randomly assigned to several experimental and control groups. Care was taken to include in the non-TM groups programs that would control for any expectation or placebo effect. Shecter found that the TM group, as compared to the control groups, showed reduced anxiety and increased energy level, self-esteem, tolerance, creativity, and intelligence.

Abrams (1972) found better performance on both short-term and long-term recall in those practicing the TM technique. MacCullum (1974) found increased creativity, as measured by the Torrance Test of Creative Thinking (verbal); in another study Travis (1979) found that subjects practicing the TM technique for five months showed a significant improvement in figural creativity tasks, in comparison with a control group.

Miskiman (1973) found increased spontaneous organization of newly learned material into taxonomic groups. Pagano and Frumkin (1977) found that TM meditators showed improvement in tonal memory, indicating an improvement in right hemispherical functioning. Finally, in several studies, the academic performance of TM program participants has been shown to improve, as measured by increasing grade point average in college students (Collier, 1973; Heaton and Orme-Johnson, 1975; and Kory and Hufnagel, 1974).

PRISON REHABILITATION PROGRAMS

A number of studies on the application of the TM technique in various rehabilitation settings have shown long-term beneficial changes (Orme-Johnson, 1981; Marcus, 1975; Siegel, 1981). By examining the area of prison rehabilitation programs in particular, we can gain a basic understanding of these effects.

The TM program has been introduced in over 25 prisons and correctional institutions (Ellis, 1979), and at least five major studies have been completed. At La Tuna Federal Penitentiary, Orme-Johnson and coworkers (1971) found that regularity of meditation was positively correlated with the percent decrease in physiological anxiety as measured by spontaneous skin resistance responses. Also, the scores of regular meditators decreased

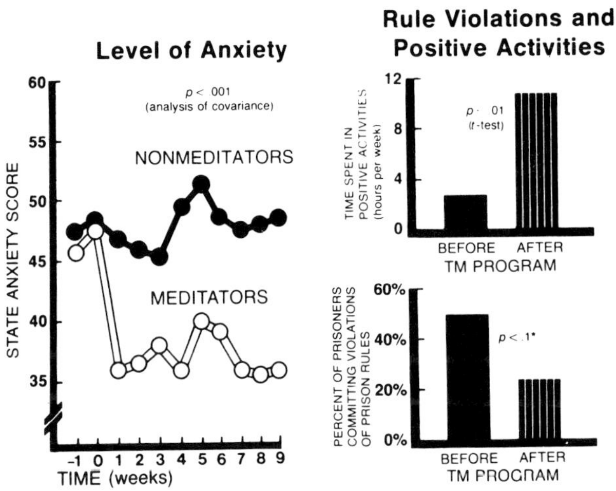

* Mann-Whitney U test comparing meditators and nonmeditators.

Finding: Three measures on prisoners practicing the Transcendental Meditation technique indicated:

1. a reduction in anxiety, as measured by the Spielberger State-Trait Anxiety Inventory (STAI)
2. a reduction in prison rule violations
3. an increase in time spent in positive activities

First Reference: David Ballou, "The Transcendental Meditation Program at Stillwater Prison," (University of Kansas, Lawrence, Kansas, U.S.A.).

Second Reference: Monte Cunningham and Walter Koch, "The Transcendental Meditation Program and Rehabilitation: A Pilot Project at the Federal Correctional Institution at Lompoc, California."

significantly, compared to controls, on several scales of the MMPI.

At Lompoc Federal Correctional Institution in California, state and trait anxiety were found to markedly decrease in inmates (Cunningham and Koch, 1973). In a study at Stillwater Prison, Minnesota, Ballou (1973a, 1973b) randomly assigned 30 volunteers to the TM technique and another 20 to a control group who were not taught the TM technique until after the experimental period. A second control group consisted of 16 inmates not desiring to begin the TM technique. The TM group decreased substantially and significantly, as compared to both control groups, on state and trait anxiety in the first weeks after instruction and maintained this level over nine weekly testings. After learning the TM technique, the meditators participated in a mean of twice as many educational and recreational activities and spent three times as many hours per week in these activities. Of 16 inmates who were subsequently paroled only one violated parole.

At the Massachusetts Correctional Institution in Walpole, Ferguson (1977, 1978) studied 57 inmates who received instruction in the TM program. A very high percentage of these inmates reported a greater ability both to deal with other people and to handle stress. In addition, Ferguson (1977) administered a number of tests, including the State-Trait Anxiety Inventory, the Buss-Durkee Hostility Inventory, as well as a sleep survey. He also studied disciplinary infractions. In the study 45 men were instructed in three groups of 15 and tested before learning TM and after three, seven, and ten weeks of TM respectively. Groups I and II showed reduced anxiety. All three groups showed reduction in several aspects of hostility-assault, indirect hostility, irritability, negativism, resentment, suspicion and verbal hostility. The sleep questionnaire showed increased deep sleep without waking. Decrease in time needed to fall asleep was also reported. Groups II and III showed reduced disciplinary infractions by 57 percent and 83 percent respectively (Group I was at a low level initially and did not change).

In a random assignment cross validation design (the same study done twice on two independent samples), which involved 115 inmates at Folsom State Prison in California, Abrams and Siegel (1978, 1979) found that the TM program significantly reduced state anxiety, trait anxiety, and neuroticism. It also reduced negativism, suspicion, assault, irritability, resentment, and verbal hostility. The meditating inmates also found that they needed less time to fall asleep and that their sleep was deeper, more continuous, and had a more restful quality.

The following letter from an inmate at Folsom gives a personal expression to the effects of the program.

> Dear Governor Brown:
>
> At the present time I am meditating in accordance with the program set up by the Science of Creative Intelligence and its founder, Maharishi Mahesh Yogi.
>
> First of all, let me tell you a little bit about myself. I feel like I have always had problems and have been in trouble all my life. Prior to 1963 I had been arrested for only checks, theft and fraud. The only treatment I had received was a short interview with a Navy psychiatrist prior to discharge. In 1963, I was sent to San Quentin with convictions for kidnapping, robbery, escape and cop shootings. During the following seven-and-a-half years, I saw a total of twenty-two psychiatrists, psychologists and counselors.

I spent hundreds of hours in group therapy, group counseling and individual therapy and was finally able to con the psychs into giving me clearance. I was subsequently released on parole in June, 1970. In November of 1970, after shooting one cop and assaulting several more, I was again arrested. Besides the above, I also picked up convictions for robberies and one conviction for first degree murder. Immediately upon my return to Folsom I became involved in smuggling and trafficking in drugs and escape attempts. All the above is true and can be verified by a telephone call to the institution.

In view of the above, it goes without saying that I was housed in the maximum security section and was going nowhere. I felt like I was buried, that I was dying. I felt like I was slowly burning out with each passing day and had nothing to look forward to but death; and with the confusion, misery, hate and frustration that I felt each day death and the release it promised started looking good to me.

...From the moment I received my TM technique to the present time, there has been a change within and it has all been positive. The confusion and frustration are gone and the hate and misery I am able to cope with easily. For the first time in my life I know where I am at, and more important, I know where I am going. I have direction! Another equally important thing is that I no longer feel the need to use narcotics. I may be living behind forty-foot high walls, but I have never felt more free in my entire life...

Governor, an extremely small number of men here are as lucky as I am in having funds available to pay the initiation fees and this is what this letter is all about. I personally feel that the TM technique is so good for the men here that I am going to pay the initiation fee for my best friend so he can receive the benefits I have.

There are literally hundreds and hundreds of men here who are aware of the TM program and want to get involved but will be unable to do so because of the lack of funds. To me this is criminal. All that the current programs being funded by C.D.C. do is teach a man to be a cross between a robot and a parrot: act right and talk right and you will be paroled. The TM program is the only program that starts inside the man's head and has immediate results. In all honesty, I can say that after twelve and one-half years in San Quentin and Folsom, with all their psychs and groups, the Transcendental Meditation program, as taught by Maharishi Mahesh Yogi, has been the only thing that has ever helped me and I know that it can help the rest of the men here.

The beneficial effects of the TM program are well documented. The tests conducted at two federal prisons, La Tuna in Texas and at Lompoc here in California, showed nothing but positive gains.

> The men here at Folsom are at the end of the line and any funds that you could make available for classes in the Transcendental Meditation program would be greatly appreciated.
>
> Sincerely yours,
>
> (signed)
>
> E. J. Corum

Unfortunately, funds were not made available at that time.

One of the most comprehensive and well-controlled studies on the effects of the TM program was done by Alexander as his doctoral thesis in psychology at Harvard University (Alexander et al., 1982). The first part of the study, conducted at Massachusetts State Male Prison, involved a double-blind cross-sectional survey of long-term TM participants and subjects interested or not interested in learning the TM technique. The three groups did not significantly differ on preincarceration variables, including age, education, race, religion, IQ, and seriousness of crime. Nor did subjects interested and not interested differ on 20 personality and developmental test variables.

Covarying for IQ, education, religious affiliation, time of testing, and months currently served, the long-term TM participants scored significantly higher on Loevinger's Ego Development Scale than the non-meditating subjects interested and not interested in TM. Three times as many non-meditators as meditators fell in the low ego development range. The long-term TM subjects also scored significantly higher on Alexander's Stage of Consciousness Inventory (SCI) and the TAT-measured Intimacy Motivation test, and significantly lower on state-trait anxiety, self-reported number of stresses, and on seven psychopathology scales of the Special Hospital's Assessment of Personality and Socialization (SHAPS), including psychopathic deviation and aggression.

Upon completion of cross-sectional testing, the group of subjects interested in TM was divided into Start TM and Delay Start TM subgroups. Results of a longitudinal follow-up (after an average of 14 intervening months) were consistent with cross-sectional findings. Notably, the Start TM subjects who were regular in the practice showed an increase in ego development of one step, which was highly significant, while the Delay Start TM and not interested groups remained unchanged. The long-term TM subjects went up a full step during the follow-up, suggesting that the TM program may induce continuous growth.

Typically, ego development "freezes" after 20 years of age. Students usually progress one step during four years of college. After reaching the age of about 26, when ego development has normally stopped, the long and short-term meditating prison inmates progressed as much during a little over one year as college students do in four. Of particular importance, the study showed a significant reduction of 20 percent in rate of recidivism in TM subjects as compared to controls and subjects involved in several different self help organizations (Alexander et al., 1982).

These findings are corroborated by another study conducted with the cooperation of the California Department of Corrections which found an overall 40 percent reduction in recidivism (revocations of parole plus new prison terms) in subjects who learned the TM technique at Folsom, DVI, and San Quentin, as compared to controls matched for offense, prior commitment record, year of birth, ethnic group, institution, and parole year. This reduction included a 56 percent reduction in new prison terms and a 31 percent increase in clean parole records by 201 parolees (Bleick and Abrams, 1982).

Bleick calculated the cost benefits if 10 percent of the annual pool of the California Department of Corrections parolees, or 1000 individuals, received instruction in the TM program. For simplicity, she assumed that all TM instruction and all paroling would take place at the beginning of each year. (Thus a year's worth of financing will be saved by the end of the year following the year in which TM instruction takes place. Under actual circumstances savings would be spread over a longer time span, but the amount would be the same.)

Bleick allowed a $10,000 cost for each revocation of parole and a $20,000 one-year cost for each new prison conviction. On the basis of average 1978 through 1980 parole year figures for the California state-wide parolee population and the one-year parole outcome figures for the California TM group, the 1000 TM parolees would save the state about $2.0 million. Bleick also arrived at national figures by assuming nationwide costs and recidivism rates to be about the same as in California. A nationwide TM program for 10 percent of the felon population would yield cumulative net savings of $170 million after five years.

Bleick, referring to the large percentage of prison inmates starting TM in certain states such as Vermont (Gore et al., 1984), suggests that it is not unreasonable to assume that up to 25 percent or more of a state prison population would voluntarily undertake TM instruction if it were made available. If TM instruction were also made available in county jails, in the juvenile authority, and in the educational systems as a whole, the overall saving could be dramatically increased. There has never been a program that offers such far-reaching advantages and concrete benefits to the field of rehabilitation as the Maharishi Technology of the Unified Field.

Reduced Criminal Recidivism

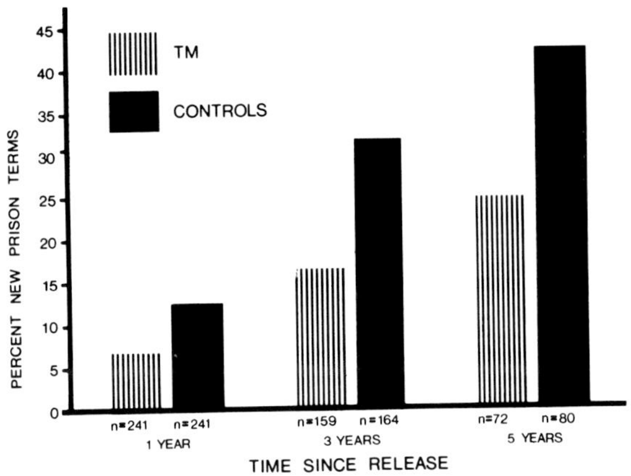

250 felons paroled from the California Department of Corrections in 1976 - 1982 showed substantially reduced recidivism after receiving TM instruction.

1) By comparison to a control group matched for parole year, institution, prior commitment record, age, race, and drug use history, the TM group had more clean records, more arrests or jail sentences under 90 days (considered by the California Department of Corrections as a relatively favorable outcome), and many fewer new prison terms at one through five years after release as determined from rap sheets ($p<.05$).

2) With other recidivism predictors (such as IQ, pre-TM prison rules violations, and participation in prison-sponsored education and psychotherapy) compensated for statistically, the TM group overall had about 25% new prison terms vs. 35% for the controls.

3) TM recidivists surveyed in prison while serving new terms or parole revocations were significantly ($p<.002$) less regular in meditation than the not-yet-released meditators.

Reference: Catherine R. Bleick and Allan I. Abrams, "The Transcendental Meditation Program and Criminal Recidivism in California," (Institute for Social Rehabilitation, USA, 1984).

SELF-ACTUALIZATION

There are a number of psychological studies on the TM program that show improvements in self actualization, as measured by tests based on the research and theories of Abraham Maslow (Ferguson and Gowan, 1976; Hjelle, 1974; Nidich et al., 1973; Seeman et al., 1972; and Shapiro, 1974). This is particularly interesting since Maslow's description of self actualization has a number of similarities to Maharishi's description of the growth of higher states of consciousness (Maslow, 1968).

According to Maslow, there are two basic tendencies or types of individuals: those he calls "deficiency motivated" and those who are "growth motivated." Maslow explains that if a person is deficiency motivated, he is able to perceive and cognize the world in a manner that is organized only by his deficiency needs. The growth-motivated person has a different type of cognition, a cognition of Being, which allows him to perceive and experience different values in the world. According to Maslow such individuals have sufficiently gratified their basic needs for safety, belongingness, love, respect, and self-esteem and are motivated by the desire for self-actualization. Among the values perceived and experienced by this individual are wholeness, unity, completeness, playfulness, honesty, self-sufficiency, and integration. These values, according to Maslow, are often perceived in so-called "peak experiences", or "transcendent experiences." Maslow feels that not having these experiences "may be a lower, lesser state, a state in which we are not fully human, not sufficiently integrated." As he says, "When we are well and healthy and adequately fulfilling the concept commonplace" (Maslow, 1969).

Maslow distinguishes two kinds of self-actualizing people: those "...with little or no experience of transcendence and those in whom transcendent experience was important—even central." Maslow feels that transcending self-actualizers are, among other things, more inclined to be responsive to beauty, more holistic in their perception of the world, more easily able to transcend their identity, more lovable and awe-inspiring, more apt to be innovators, more apt to have a strong self-identity and self-knowledge, more open and humble, and more integrated and self-confident. These characteristics delineated by Maslow are remarkably parallel to Maharishi's description of the characteristics of individuals living higher states of consciousness.

The very wide range of physiological and psychological effects seen in TM subjects outside of meditation strongly suggests that the continued practice of the TM technique results in a developmental process in which there are improvements in all aspects of the individual's physical, mental, and emotional capacities. By enabling the nervous system to support the least excited state of consciousness, the state of pure consciousness, the Maharishi Technology of the Unified Field cultures a new style of neurophysiological functioning in which human awareness is brought in tune with the unified field of natural law.

As a result of frequent and regular experience of this field, pure consciousness begins to be experienced along with the other three states of consciousness, and all levels of physical and mental activity begin to function in accord with and enjoy the support of the full potential of natural law in a very spontaneous manner, and life rises naturally to higher states of consciousness. This process is further accelerated with the more advanced TM-Sidhi techniques, and results in a wide variety of specific long-term changes in physiological functioning. The continued progress of research, as we shall see, has enabled a far more detailed understanding of the neurophysiological processes which underlie the development of higher states of consciousness.

Chapter 5

RESEARCH ON THE TM-SIDHI PROGRAM

The Transcendental Meditation technique has been described as a procedure that opens one's awareness to the state of pure consciousness, the least excited state of consciousness, the ground state or unified field of natural law. In terms of the physiology, this process involves the gradual refinement of neurophysiological functioning such that the nervous system becomes capable of supporting a state of "restful alertness." In this state, the activity of autonomic and metabolic functions become minimal, while the electrophysiological activity of the brain adopts a highly coherent style of functioning.

The TM-Sidhi procedures take this development and refinement of neurophysiological functioning a step further and reveal how, by creating impulses of activity in the silent nature of pure awareness, the least excited state of consciousness can be stabilized in human awareness. According to Maharishi, the stabilization of the total potential of natural law in human awareness results in the ability to think and act spontaneously in accord with all the laws of nature.[1]

The principle underlying the TM-Sidhi program is that

---

1. A more detailed explanation of the theoretical basis for how the TM and TM-Sidhi program makes use of the total potential of natural law, is given in chapter 9.

pure consciousness, the simplest form of human awareness, is the underlying unified field that structures not only thought processes but also the physical laws of nature—it is literally a field of all possibilities. The TM-Sidhi program, as we have mentioned, originates in the Vedic tradition of India, in particular in the Yoga Sutras of Patanjali. The basic element in the TM-Sidhi program is the experience of "sanyama," the procedure through which specific sutras or mental formulas are used to gain specific results.

Sanyama consists of three values: dhyana, dharana, and samadhi. Dhyana is the movement of the mind towards the transcendent, dharana is the fixation of the mind on a specific mental impulse (a particular sutra), and samadhi is the transcendent. When the two values of dharana and dhyana come together at the point of samadhi, the result of the specific sutra is achieved. Patanjali describes sanyama as a spontaneous mental process that begins to occur once the state of samadhi, pure awareness, is established to a sufficient degree of stability that it can simultaneously coexist with faint impulses of mental activity. Then, Maharishi says, it becomes possible for pure consciousness, the ground state of natural law, to "adopt" a thought or intention in such a way that it is spontaneously carried out without any effort or action on the part of the individual.

By enlivening specific impulses—laws of nature—at the level of the unified field, one gains the ability to create powerful and positive effects in one's own nervous system and in the environment by the mere intention. The TM-Sidhi program develops specific abilities of the human nervous system—perhaps the most striking example so far is the "flying" sidhi—but its main purpose and end result is to develop complete mastery over natural law.

Maharishi explains that the TM-Sidhi abilities ("sidhi"

literally means "perfection") are not to be regarded as "supernormal," but as normal human abilities unfolding. In terms of neurophysiology, they could therefore be considered as species-specific traits that are encoded in the genetic potential of the human nervous system. They require, however, a certain level of stabilization of neuronal coherence and activation before the pre-programmed neuronal information can spontaneously be read out. The successful performance of the TM-Sidhi program is not specific to any race or culture but is rather an inherent capacity of the human nervous system. The traditions of many cultures contain accounts in which certain individuals at various times in history appear to have made use of these abilities.

In this chapter we will examine the research on the physiological and psychological effects of the TM-Sidhi program. We will emphasize, in particular the electrophysiological research, which indicates the ability of these procedures to optimize neurophysiological functioning. The more far-reaching sociological effects of the TM-Sidhi program will be examined in Chapter 8.

EEG COHERENCE

The first studies conducted on participants in the TM-Sidhi program were by Orme-Johnson and coworkers (1977), and Orme-Johnson and Haynes (1981). As we briefly mentioned earlier, Orme-Johnson's group studied twelve subjects who reported clear experiences of transcendental consciousness during the TM technique, clear experiences of transcendental consciousness during a night's sleep, and clear experiences of the specific sutras during the practice of the TM-Sidhi program. He found that these subjects exhibited EEG coherence in the alpha and theta frequencies in the frontal and central areas of the brain, compared to ten male subjects whose experiences were less clear.

Experiences of the Sidhis and Increased Brain Wave Coherence

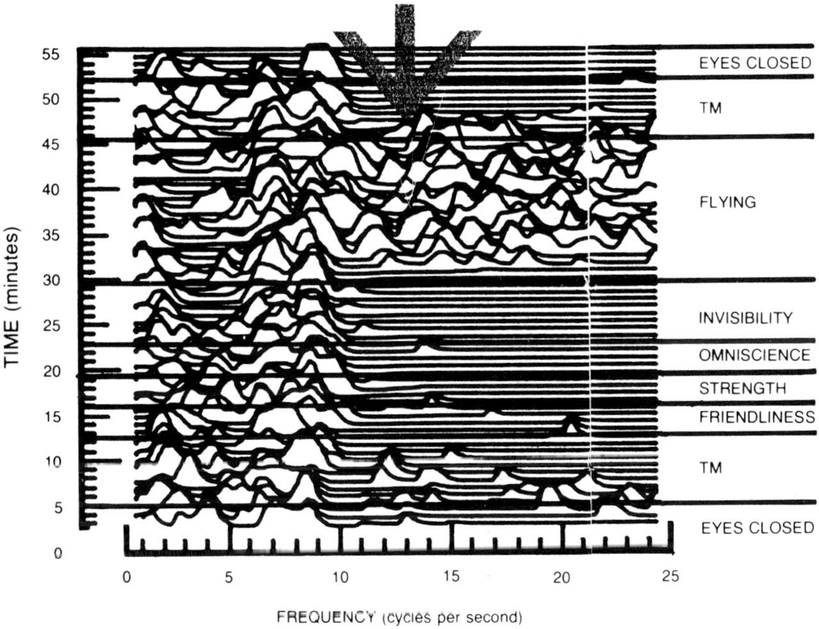

ORME-JOHNSON, D. W.; CLEMENTS, G.; HAYNES, C. T.; and BADAOUI, K. Higher states of consciousness: EEG coherence, creativity, and experiences of the sidhis. Centre for the Study of Higher States of Consciousness, Maharishi European Research University, Switzerland, 1977.

INCREASED EEG COHERENCE DURING PRACTICE OF THE FLYING SIDHI. Consistent increases in EEG coherence for a subject during practice of the flying sidhi.

High Brain Wave Coherence, Stabilized Pure Awareness, Increased Creativity, and Experiences of the Sidhis

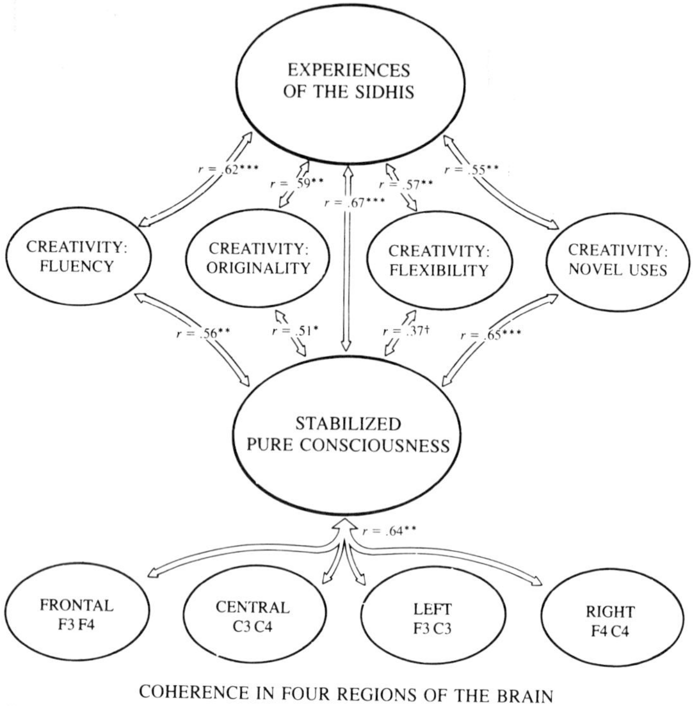

*** = $p < .005$
** = $p < .01$
* = $p < .05$
† = N.S.

Probability values are for one-tailed tests. Correlations with experiences of sidhis are for N = 22. Correlations with stabilization of pure consciousness are for all subjects for which data were available (N = 15).

CORRELATIONS AMONG EEG COHERENCE, STABILIZED PURE CONSCIOUSNESS, CREATIVITY, AND EXPERIENCES OF THE SIDHIS. Significant positive correlations were found between the abundance of alpha EEG coherence in four regions of the brain and stabilized pure awareness (transcendental consciousness) and among stabilized pure awareness, creativity, and frequency of experiences of the sidhis.

ORME-JOHNSON, D. W.; CLEMENTS, G.; HAYNES, C. T.; BADAOUI, K. 1977. Higher states of consciousness: EEG coherence, creativity, and experiences of the sidhis.

The areas of the brain that showed the highest amount of EEG coherence varied considerably among the individuals in both groups. There was, however, a significant correlation between degree of alpha coherence (in whichever area of the brain showed highest coherence) and the clarity of the experience of transcendental consciousness during the TM technique. The subjectively reported experience of transcendental consciousness during sleep (referred to as "witnessing" during sleep) was also significantly correlated with alpha coherence over all EEG electrodes, and less strongly with theta coherence and beta coherence. Further, all subjects showed a positive correlation between the following variables: the various subscales of the Torrance Test of Creative Thinking (fluency, originality, flexibility, and novel use), total EEG coherence, experience of transcendental consciousness during sleep, and number of clear experiences of the sidhis. It was also found that clarity of experience of transcendental consciousness during the TM technique was strongly correlated, as would be expected, with experience of transcendental consciousness during sleep and number of clear experiences of the sidhis.

In another experiment, Orme-Johnson and coworkers made a more detailed analysis and classification of the subjective stages of development in the TM-Sidhi experience. EEG measurements made on 10 subjects while they practiced the advanced TM-Sidhi techniques showed a general increase in coherence correlated with experiences of the sidhis. One striking example showed a marked increase in the right hemisphere of the brain (C4) during the "flying" sidhi, particularly in the beta frequency range. In another subject, EEG coherence and heart rate were recorded simultaneously during the "flying" sidhi. Coherence and heart rate during the TM technique have usually been found to be negatively related. In this case, particularly during the period corresponding to a gradual "hop"[2] by the subject as observed on a TV monitor, the EEG coherence increased markedly along with an increase in heart rate. Orme-Johnson suggests that as a result of the TM-Sidhi program the subject had a greater ability to maintain the experience of pure consciousness (as shown by increased EEG coherence) even during physical activity.

One further subject whose records were analyzed in detail showed cycles of coherence in four separate areas of the brain during the TM and TM-Sidhi program. The subject was asked to press a button after a clear experience of a particular sutra. In general, the subject described his experience as a sequence: an abstract experience of pure consciousness, then a blissful sensation as the subjective experience of the specific sutra developed, followed by the thought to push the button. Orme-Johnson found that periods of high coherence occurred in the alpha and/or theta frequencies in all areas of the brain starting from approximately 5 to 25 seconds before pressing the button in 12 out of the 13 button presses. Periods of high coherence were also found to precede "hopping" in this subject. Periods of coherence were also found to correspond to slower respiration and greater stability of skin resistance, as has been noted by other researchers (Farrow and Hebert, 1982; Badawi et al., 1984).

In a longitudinal study on EEG coherence, a number of changes were seen in college student's practicing the more advanced TM-Sidhi program as compared to those practicing the TM program alone (Orme-Johnson et al., 1981, Nidich et al., 1983). Out of a subject pool of 143 freshman and sophomores who were available for testing

---

2. One of the first stages in the performance of this sutra as delineated by Patanjali.

Increased Mind-Body Co-ordination

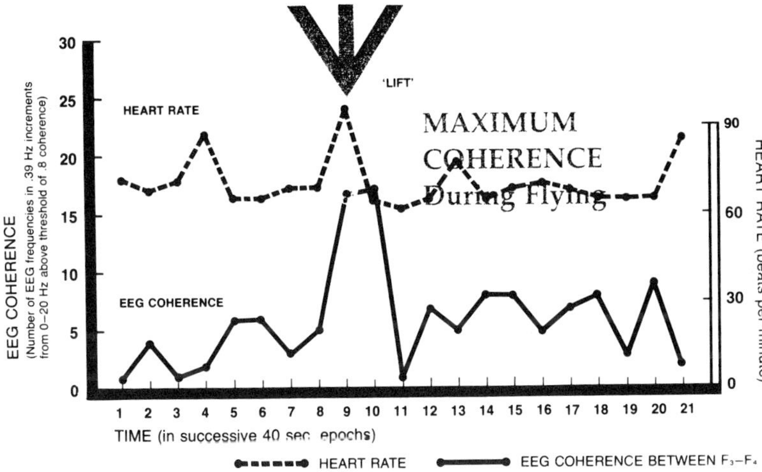

HIGH EEG COHERENCE AND HEART RATE DURING AN EXPERIENCE OF LEVITATION. The subject in this experiment had a distinct experience of lifting from the ground a small distance during the tenth second of the ninth 40-second epoch of the period of practice of the flying sidhi. It can be seen from the figure above that the "lift" occurred during a period of high bilateral frontal EEG coherence which remained at a high level during the experience. During the time of the "lift" the heart rate accelerated from 69 to 96 beats per minute. It can be seen in the figure that at other times there was generally a negative correlation between heart rate and coherence, lower heart rate corresponding to higher coherence.

Reference: David W. Orme-Johnson, Geoffrey Clements, Christopher T. Haynes, and Kheireddine Badaoui, 'Higher States of Consciousness: EEG Coherence, Creativity, and Experiences of the Sidhis,' MERU Report 7701, Centre for the Study of Higher States of Consciousness, Maharishi European Research University, Seelisberg, Switzerland, 1977. Abstract in *Scientific Research on the Transcendental Meditation Program: Collected Papers*, Vol. 1, MERU Press.

over the period of a year, and who had no previous experience in the TM-Sidhi program, approximately 24 matched pairs were randomly selected. Subjects were matched as closely as possible for age, sex, and length of time meditating. Both experimental and control groups were measured on a variety of physiological and psychological variables before and after a three-month summer recess. During this period the experimental group learned the TM-Sidhi program while the control group took their usual summer vacation.

A multivariate analysis showed a significantly greater increase in alpha and theta EEG coherence outside of meditation during resting in the TM-Sidhi subjects than in the control subjects. An analysis of the alpha index (derived from calculating the percentage of alpha wave activity) showed no significant difference from pre to post testing, suggesting that EEG coherence provides a unique type of information not available by classical EEG scoring methods. The most striking changes in EEG coherence were seen in the frontal and central portions of the brain, as compared to the occipital or posterior portions of the brain. This finding is of particular interest since Orme-Johnson and Levine have shown that frontal alpha coherence seems to be a better indicator of wakefulness than occipital coherence. Even during brief periods of drowsiness, frontal alpha coherence markedly decreases. Thus the findings suggest that the state of increased inner alertness or wakefulness which is seen during the TM and TM-Sidhi program may result in a long-term trait effect of greater inner alertness outside of meditation.

Finally, a significant correlation was found between alpha frontal coherence and a variety of performance measures, including IQ (Raven Progressive Matrices), creativity (Torrance Test of Creative Thinking), grade point average, scores on the Scholastic Aptitude Test, and moral judgment (Kohlberg's Moral Development Scale). The positive correlation of frontal alpha EEG coherence with intellectual and moral development further reinforces this interpretation of a more alert and optimal style of brain functioning.

Traditionally, the process of gaining enlightenment is referred to as a procedure of becoming more and more fully awake inside, awake to the inner dynamics of consciousness. According to Maharishi, to be "fully awake" means to be established in the state of pure awareness, the unified field of natural law. Established in that state the individual performs optimally: he acts in accordance with all the laws of nature. The poet Thoreau also recognized the value of being fully awake, and in the following quotation beautifully summarizes the importance of this state:

> Moral reform is the effort to throw off sleep. Why is it that men give so poor an account of their day if they have not been slumbering? Had they not been overcome with drowsiness they would have performed something. The millions are awake enough for physical labor; but one in a million is awake enough for effective intellectual exertion, only one in a hundred million for a poetic or divine life. To be awake is to be alive. I have never yet met a man who was quite awake. How could I have looked him in the face?

SENSORY AND MOTOR RESPONSES

Several studies have been conducted on the effects of the TM-Sidhi program on sensory and motor performance. Clements and Milstein (1977) determined absolute hearing thresholds by using pure sine wave tones from 250 Hz to 8000 Hz in eight female subjects participating in the TM-Sidhi program for approximately eight months.

Absolute Auditory Threshold

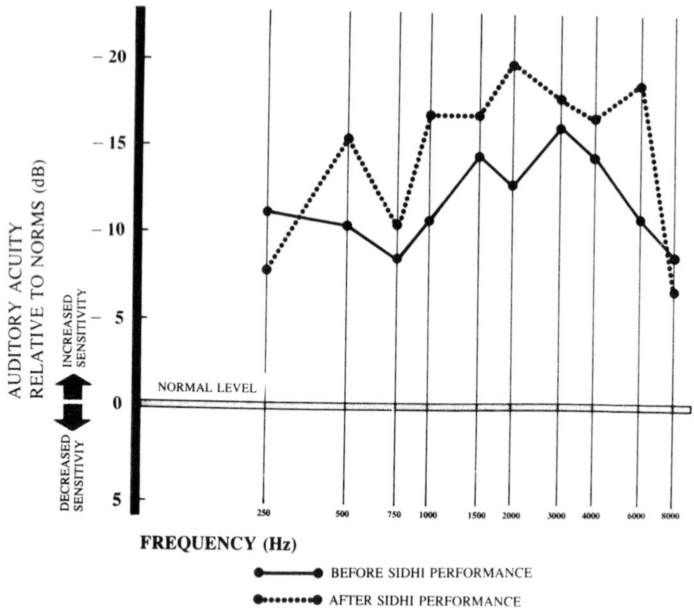

ENHANCED HEARING SENSITIVITY IN ADVANCED PARTICIPANTS IN THE TRANSCENDENTAL MEDITATION PROGRAM. The absolute threshold of hearing at frequencies from 250 Hz to 8000 Hz was determined using a standard von Bekesy technique for eight female students (ages 23–30) participating in an advanced Age of Enlightenment Governor Training Course at MERU. The subjects displayed unusually sensitive hearing thresholds (11.7 dB more sensitive than norms averaged over all frequencies). After 15 minutes of performing the TM-Sidhi technique to enhance hearing ability, hearing became even more acute by a further 3.0 dB.

CLEMENTS, G., and MILSTEIN, S. L. Auditory thresholds in advanced participants in the Transcendental Meditation program. Department of Experimental Psychology, Maharishi European Research University, Switzerland, and Institut National de la Recherche Scientifique, Centre de Recherche en Sciences de la Santé, Université du Québec, Montréal-Gamelin, Québec, Canada, 1977.

The subjects in the study had been performing a number of TM-Sidhi procedures designed to enhance sensory abilities, including one that has the specific intention of enhancing hearing beyond the usual range of sensitivity. The auditory thresholds were determined before and directly after a 20-minute period consisting of five minutes of the TM technique followed by 15 minutes of performing the sutra to enhance hearing ability.

The results showed unusually low hearing thresholds for almost all subjects (11.7 decibels more sensitive than norms). The hearing threshold decreased significantly (a further 3.0 db) after 15 minutes of performing the TM-Sidhi designed to produce enhanced hearing ability. The authors examined subjective reports of practitioners and noted that, after the performance of that sidhi, subjects commonly reported their hearing was extremely clear. As one subject reported, "Hearing is becoming more and more acute--much fuller and very sweet and blissful. All senses are more refined, not necessarily just sharper, but fuller, with intuition and knowledge of greater depth."

The authors suggest that the improvements in hearing ability might be due to a reduction of noise within central auditory pathways. Previous findings (Levine et al., 1975; Orme-Johnson et al., 1977a) of higher EEG coherence during the TM and TM-Sidhi program suggest higher stability of EEG activity and therefore lower internal noise levels within the cortex.

McEvoy and coworkers (1980) studied brainstem auditory evoked potentials in TM-Sidhi program participants. Using a design similar to that of Clements and Milstein, they found a significant change in brainstem auditory evoked potentials after the practice of the sidhi designed to enhance hearing, as compared to before the practice. They also suggest that their findings indicate a reduced signal-to-noise ratio in the central processing of auditory information.

Orme-Johnson and Granieri (1977) showed longitudinal improvements in field independence in students, which, as we noted in the previous chapter, indicate a more stable and holistic perceptual ability. The students were able to complete the tests as much as 200 percent faster than they were designed for, necessitating major adjustments to prevent a "ceiling effect."

In the area of motor performance, several studies have focused on the motor reflex responses as measured by the paired Hoffman or H-reflex. The H-reflex was first described by Hoffman in 1918 and has since been extensively studied under a variety of experimental and pathological conditions (Hugon, 1973). In the H-reflex the sensory nerve from the calf muscle is excited by passing an electrical stimulus over a small area of skin below the knee cap, where the nerve is partially exposed on its way from the muscle to the spinal cord. The stimulus mimics the action of a sensory receptor and, via a monosynaptic reflex, excites the motor neuron in the spinal cord and causes the calf muscle to contract.

The electrical potential of a muscle contracting, called the H-wave, can be recorded and measured. The paired H-reflex involves a slight modification of the original procedures. Instead of a single stimulus, pairs of stimuli, separated by varying time intervals, are given. In general the magnitude of H-wave following the second stimulus is less than that following the first. By comparing the magnitude of each of the H-waves elicited from the paired stimuli and by varying the time intervals between, an H-reflex recovery curve is obtained. The H-reflex recovery curve is thus a measure of the recovery of the H-reflex after it has been conditioned by a preceding stimulus.

Haynes and coworkers (1976) reported that subjects with clear experience of pure consciousness showed faster recovery of the paired H-reflex; and in addition, faster recovery of the paired H-reflex response correlated significantly with higher EEG coherence and creativity measures. More recently, Dillbeck and coworkers (1981a) reported a significant correlation between flexible performance on a concept learning task and faster recovery of the paired H-reflex response and frontal EEG coherence in TM-Sidhi participants.

In a study in our laboratory at MIU we investigated the effects of the TM-Sidhi program on paired H-reflex recovery at nine delay intervals (50, 70, 100, 150, 200, 250, 333, 500, and 1000 msec) through a longitudinal study (Wallace et al., 1983). An experimental group of 14 mentally and physically healthy participants in the TM program were matched with eight healthy controls also practicing the TM program. The experimental group was instructed in the TM-Sidhi program while the controls practiced TM only. The subjects were tested before and after the same three-month period. Significant longitudinal differences were found in H-reflex recovery between the male TM subjects and the male TM-Sidhi subjects, between the time intervals 100 msec and 250 msec, the period most intensively studied.

Several researchers have studied the relationship of H-reflex recovery to different states of awareness. For example, Pivik and Mercier (1979) showed that during non-rapid eye movement (NREM) sleep (stages 2 and 4) the H-reflex recovery during the period between 100-300 msec was significantly reduced. Van Boxtel (1976) made simultaneous measurements of EEG alpha activity and H-reflex recovery curve. He showed that a constant alpha index (percent of time alpha was present) was accompanied by stable H-reflex amplitudes and that a decreasing alpha index (indicative of drowsiness) was accompanied by decreasing H-reflex amplitudes.

These studies suggest the involvement of a system in the brain known as the reticular formation, which is concerned with both maintaining the tone of consciousness and with regulating reflex activity. The correlation of the H-reflex recovery response in TM-Sidhi participants with higher levels of EEG coherence indicates the type of neurophysiological mechanism suggested by both Pivik and Mercier and Van Boxtel. As we shall discuss in Chapter 9, the reticular formation (in particular, the reticular activating system) has been suggested by a number of researchers to play an important role in the development of the state of inner awareness in higher states of consciousness.

Studies on TM subjects have also shown a significant correlation between faster recovery of the paired H-reflex response and higher academic performance, thus suggesting a general effect of enhanced inner wakefulness as a result of the TM-Sidhi program (Wallace et al., 1982b). While the mechanisms for altering the H-reflex recovery curve remain complex, in this as well as in other reflex studies the H-reflex provides an interesting model for the study of the development of higher states of consciousness.

BIOCHEMICAL STUDIES

A comprehensive longitudinal study on plasma hormone levels was recently reported by Werner and coworkers (1986) on TM-Sidhi participants. Blood samples were taken in subjects practicing the TM technique just prior to their starting the more advanced TM-Sidhi program, and then during sessions at intervals of 5, 49, 115, and 167 weeks after starting. During each session samples were taken over five consecutive days. The results showed significant longitudinal decreases in some hormone levels, particularly in the pituitary hormones, thyroid stimulating hormone (TSH), growth hormone, and prolactin, with no marked change in cortisol or the thyroid hormones T3 and T4.

One of the most interesting findings of the study, the

relatively unchanged levels of the thyroid hormones (despite continually decreasing TSH levels), implies either a marked increase in the sensitivity of the regulation of the thyroid gland, or the production of more biologically active TSH. The latter is an interesting hypothesis since it is known that TSH levels are elevated in elderly subjects. This interpretation would be consistent with the other findings discussed in a later chapter which suggest a reversal of the aging process in TM subjects.

Another important finding was the simultaneous longitudinal decrease in the three pituitary hormones measured—TSH, growth hormone, and prolactin—which suggests that the TM-Sidhi program affects the regulation of anterior pituitary function. Mechanisms known to be important in the control of anterior pituitary function include such factors as: neurotransmitter levels in the hypothalamus, hypothalamic releasing factors, and pineal function. It is not possible from this study to distinguish which of these mechanisms might be influenced by the long-term practice of the TM-Sidhi program. However, the simultaneous decrease in all three hormones suggests a global mechanism. It may be that the TM-Sidhi program increases melatonin secretion by the pineal gland, which then decreases the secretion of the pituitary hormones TSH, prolactin, and growth hormones either directly or indirectly by affecting transmitter levels in the hypothalamus.

The findings of the above study reveal one further longitudinal effect of the TM-Sidhi program on endocrine stability. Several of the hormones exhibited reduced day-to-day variation within measurement sessions after the subjects had been practicing the TM-Sidhi program for two years. This finding, which suggests reduced ultradian fluctuation, is particularly interesting because it indicates a unique and more stable style of neuroendocrine functioning.

Longitudinal Endocrine Changes

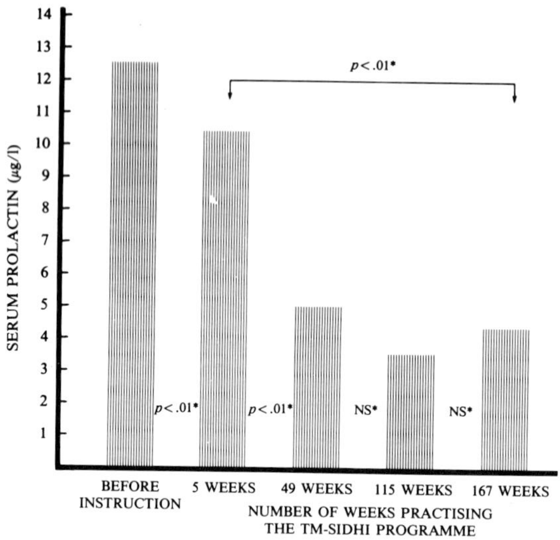

* Newman-Keuls' test (between consecutive measurements unless otherwise shown).

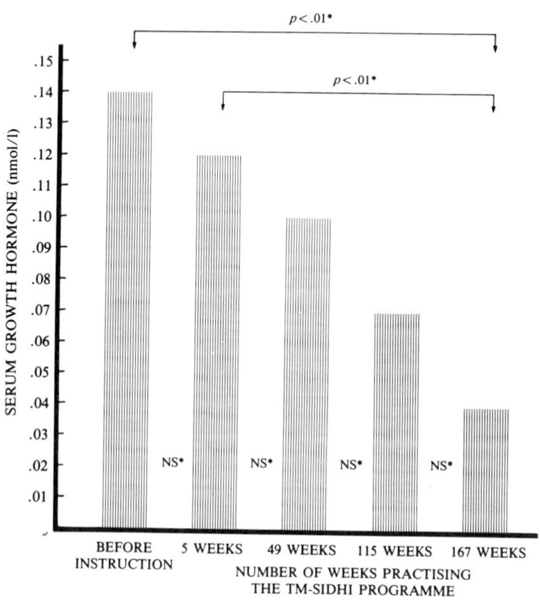

* Newman-Keuls' test (between consecutive measurements unless otherwise shown).

Longitudinal Endocrine Changes

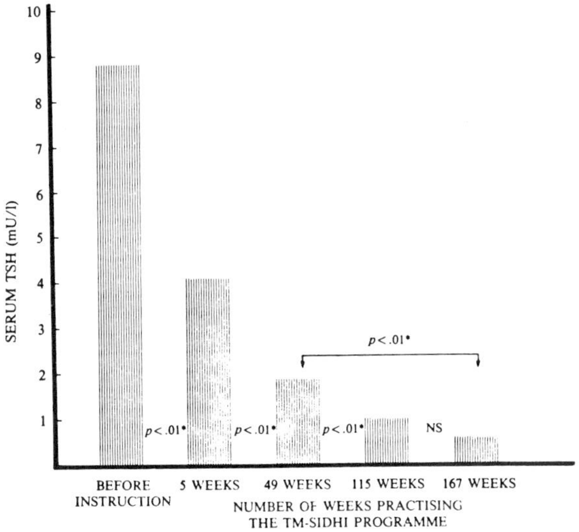

• Newman-Keuls' test (between consecutive measurements unless otherwise shown).

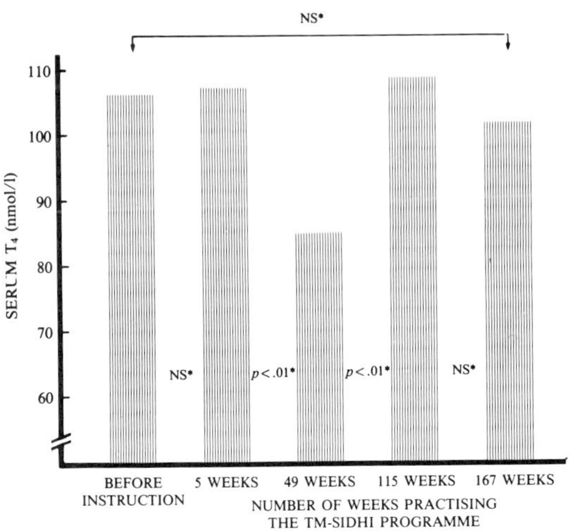

• Newman-Keuls' test (between consecutive measurements unless otherwise shown).

Werner, O.; Wallace, R. K.; Charles, B.; Janssen, G.; Stryker, T.; and Chalmers, R. Long-term endocrinologic changes in subjects practicing the Transcendental Meditation and TM-Sidhi program. *Psychosomatic Medicine*, 48(1–2): 59–66, Jan.–Feb. 1986.

One of the most exciting and promising biochemical studies on the development of higher states of consciousness is the recent investigation into an unknown substance by Walton and coworkers (1986). These investigations originated from the earlier studies of Walton and coworkers in which they were measuring urinary levels of the serotonin metabolite 5-hydroxy-indole acetic acid (5-HIAA) in subjects practicing the TM and TM-Sidhi program.

Using a general colorimetric assay their measurements suggested circadian rhythms as well as a specific increase in 5-HIAA after the practice of the TM-Sidhi program. However, when they used a more specific colorimetric assay for 5-HIAA, the same subjects did not show the same changes as previously seen. They therefore concluded that part of the substance being measured in the more general assay was not 5-HIAA, but some unknown substance which the more specific assay did not detect. They then attempted to partially isolate and purify this unknown substance.

Prelimanary evidence suggested that this unknown substance had a high affinity for imipramine receptors (imipramine is an anti-depressant) and that it blocked serotonin uptake in platelets. More recently, the substance, referred to as "substance M," was also partially isolated from blood samples. Studies are currently underway using high pressure liquid chromatography and mass spectrophotometry to purify and isolate substance M. If, indeed, substance M is a new biochemical substance produced as a result of the TM and TM-Sidhi program and if it is responsible for mediating many of the wide variety of physiological changes seen both during and after the TM-Sidhi program, then this would be a highly significant scientific discovery.

In this regard it is interesting to note Maharishi's explanation that the Vedic literature refers to a particular substance called "soma," which is associated with the state of pure consciousness and the establishment of enlightenment:

> ...a normally functioning nervous system, free from stress and strain and any abnormality, produces a chemical called soma.... If there are no restrictions, no inhibitions, then awareness is unbounded; and when this unbounded awareness is maintained spontaneously at all times, then the nervous system is functioning normally.... Now, the best product of such a normally functioning digestive system is soma.... So soma is that which helps all the fundamentals of individual consciousness rise above boundaries, and have an unbounded status. Another thing is that in that unbounded self-awareness the perception is very rich—the perception is richest!

The discovery of such a substance, unique to the experience of unbounded consciousness, would be perfectly in line with the current progress of neurochemical research, which has attempted to describe all mental states in terms of the biochemistry of the brain. This discovery would have important implications for our understanding of the relationship between the brain and consciousness.

CONCLUSION

The research on the TM-Sidhi program contributes considerably to our previous understanding of the short and long-term effects of the TM program. The TM-Sidhi program not only offers a more advanced technology to study the physiological effects of the growth of higher states of consciousness but more importantly, it also provides the practical means to explore the full range of possibilities for human awareness and its ability to influence the environment. The ability to successfully perform the TM-Sidhi techniques constitutes a highly standardized and reliable test of the growth of consciousness, a physiological indicator of the development of mind-body coordination and the ability to act in accordance with natural law. The TM-Sidhi program is thus the most advanced area of research in higher states of consciousness, resulting in the uncovering of new levels of inner experience, each of which can be verified by reference to the Vedic literature.

The Yoga Sutras of Patanjali have existed for centuries as precise scientific formulas for the development of consciousness. However, lack of knowledge of how to use them properly, and lack of experience of transcendental consciousness, which is the basis for their successful performance, made them inaccessible and incapable of producing the desired results. Maharishi's revival of the wisdom of Patanjali in its completeness has led to a breakthrough in human development--the ability to gain mastery over natural law.

The first stage of experience provided by the TM program is the fourth state of consciousness, transcendental consciousness. The original TM research scientifically verified the fourth state of consciousness as a state of restful alertness.

The studies on the TM-Sidhi program have been a breakthrough in TM research. With the introduction of the TM-Sidhi program, the silence of transcendental consciousness became active, dynamic. The field of Patanjali is the field of transcendental awareness becoming fully awake in its own nature, resulting finally in the value that Patanjali called "kaivalya," the singularity of pure consciousness as an established reality.

As TM meditators gain increasing familiarity with this field of transcendental consciousness, they are developing an appreciation of the very texture and fabric of transcendental consciousness, to the extent that they are now beginning, through the TM-Sidhi program, to distinguish the different individual impulses of natural law within the unmanifest wholeness of the unified field. This reflects an extraordinarily delicate level of inner perception. It seems that the time is now ripe to begin to investigate an entirely different system of Vedic technology, the body of sutras attributed to Jaimini, which may be described as that field of investigation which locates the unmanifest activity within pure silence. The system of Jaimini is foreseen as the next source of breakthrough in research in the Transcendental Meditation program.

The final level of development in the TM program will undoubtedly involve the last and highest system of Vedic knowledge, the system of Vedanta. In the Brahma Sutras of Vyasa lies the technology for the unification of the values of silence and activity into a single field which can exist only on the level of pure knowledge, knowledge so complete as to deserve the name Brahman consciousness—complete wholeness of awareness.

This research is of a unique kind, involving a stage-by-stage rediscovery of ancient wisdom. These discoveries are not, by the way, purely those of India, any more than the laws of electromagnetism belong solely to Germany or England. As scientists we have only begun to verify the practical benefits of the Maharishi Technology of the Unified Field, but already the evidence has revealed a new frontier for human development.

Chapter 6

PERFECT HEALTH—LIFE IN ACCORDANCE WITH NATURAL LAW

> Health means wholeness. The Sanskrit word for health is "Swasthya." "Swa" means Self and "sthya" means established. So health is defined as being established in the Self. The Self is holistic in nature. Being established in the Self means being established unshakably in the wholeness of life. Health, therefore, means possession of the Self. He who possesses the Self is described as one who is healthy.
>
> — Maharishi

How is it possible to prevent disease? According to Maharishi, all sickness arises from the violation of natural law. Perfect health means wholeness on all levels of life. When individual life is established in the unified field of all the laws of nature—the Self—all actions are spontaneously in accord with natural law. In terms of physiological activity this means perfect integration and balance, from the finest level of molecular functioning to the grossest levels of physiological organization. In examining the evolution of life physiologists recognize that greater internal balance is achieved through the development of more refined homeostatic mechanisms. This enables higher organisms to be more adaptable and therefore freer from the deleterious effects of living in a changing environment.

What is the ultimate level of balance and integration? Is it possible to develop optimal functioning of human homeostatic processes? According to Maharishi, the precise mechanics of how to achieve perfect physiological balance, and thus perfect health, are included in one specific area of the Vedic literature known as Ayurveda.

AYURVEDA: THE SCIENCE OF LIFE

Ayur means life, life lived for its full value, life in wholeness. Veda means knowledge. Therefore Ayurveda means knowledge or science of life.

Maharishi explains that Ayurveda deals with how to restore balance in the physiology and in consciousness. Every state of consciousness is associated with a specific corresponding physiological state. Abnormalities in awareness may be detected through their effects both in the behavior and in the physiology. Further, these abnormalities may be treated through four basic approaches: on the level of consciousness, physiology, behavior and the environment.

The Ayurvedic approach on the level of consciousness is the Transcendental Meditation and TM-Sidhi program. The ancient records of Ayurveda specifically refer to gaining perfect health through the practice of meditation and the practice of the sidhis as described in Patanjali's Yoga Sutras. Unfortunately, without the knowledge of higher states of consciousness and an effective technology of meditation, for many centuries Ayurveda has not emphasized the development of consciousness as an approach to perfect health. With Maharishi's introduction of Vedic Science and the Maharishi Technology of the Unified Field, Ayurveda is now being revitalized and restored in its completeness. In this chapter we will consider some of the fundamental principles of Ayurveda. First, however, let us first review the research on the effects of the TM and TM-Sidhi program on health.

THE RELATIONSHIP OF MIND AND BODY IN HEALTH

More and more researchers are realizing the importance of influencing health through the mind. In our era we find an epidemic of stress-related diseases, which clearly demonstrate the relationship between poor mental states and the increasing occurrence of disease. Stress-related diseases are by far the largest killers today, cardiovascular disease being the most prevalent. Cardiovascular disease is responsible for well over half the deaths in most technologically advanced countries.

Recent evidence has further linked various mental disorders, such as depression, and stress-related experiences with higher incidence of cancer, the second most common cause of death. Some studies suggest that mental stress either directly or indirectly may decrease the responsiveness of the immune system. While the precise mechanisms for the relationship between mental states and cancer are not known, a great deal of research is currently underway exploring the interconnections between the nervous system and the immune system (Marx, 1985).

The immune system is the major defense system of the body and is responsible for eliminating both external invaders, such as bacteria and viruses, and internal abnormalities, such as cancer. The latest research shows that the nervous system influences the activity of the immune system both directly, through nerve fibers, and indirectly, through hormones. Conversely, the immune system also seems to release certain factors in the blood that influence the activity of the nervous system. These findings represent one of the many new areas of research revealing the intimate relationship between mind and body.

From these and many other studies, particularly those concerned with the neurochemistry of behavior, it is clear that the activity of the mind influences the activity of the nervous system, this in turn causes the release of neurotransmitters and neuropeptides, along with a number of specific physiological changes which then may act back on the nervous system and influence the state of mind. Therefore, if the mind experiences more refined states of consciousness, we should be able directly to influence the style of our own neurophysiological functioning and thus our mental and physical health. As we shall see in examining the findings on the Maharishi Technology of the Unified Field, this simple mental technique improves what we might call "mental hygiene," and as a result enables the physiology to counteract or better adapt to the harmful effects of stress. Improvements have been seen in various clinical diseases. This technology, however, has even greater value as a preventive measure, a kind of "vaccine" against stress. We can clearly see these preventive effects if we examine the research on the various cardiovascular risk factors.

CARDIOVASCULAR RISK FACTORS

Cardiovascular risk factors are simply those physiological and behavioural measurements that best enable physicians to predict the probable occurence of heart disease. Some of the most important risk factors which have been highly correlated with the occurrence of cardiovascular disease are: (1) hypertension or high blood pressure, (2) hyperlipidemia or high levels of cholesterol combined with very low density lipoproteins, (3) cigarette smoking, (4) obesity, and (5) personality type.

Over one-third of the adult population of the United States, about 55 million people, have high blood pressure; about 90 percent of the cases of high blood pressure are reported as essential hypertension, a condition for which the precise cause is unknown. Many researchers feel the unknown cause of essential hypertension is nothing other than the fast pace of life today and the body's often inappropriate and prolonged physiological response to it. Physiological response to stress involves a mobilization of one's energy resources and an increase in blood pressure and blood flow to the heart, brain and skeletal muscles.

It has been suggested that the body adapts to the continual elicitation of the stress response over long periods of time by a prolonged elevation of blood pressure. In other words, the centers in the brain which regulate blood pressure become habituated to this stressed level of functioning, and eventually accept high blood pressure as being normal. A vicious cycle arises in which the prolonged high blood pressure strains the arterial system and contributes to complicating factors that further elevate blood pressure.

One of these complicating factors is hardening of the arteries, a condition known as arteriosclerosis. In arteriosclerosis, fatty substances build up on the walls of the arteries, gradually decreasing their diameter and flexibility. High blood pressure, by straining the arteries and perhaps causing internal damage to them, contributes to the development of arteriosclerosis. Once the passageway has been narrowed, the blood pressure increases still more. The narrowed arteries also increase the likelihood of a blood clot becoming lodged there and thus causing a heart attack.

There is little question about the importance of eliminating high blood pressure. Under normal clinical care, subjects are given a variety of drugs, depending upon the severity of the case, in order to reduce the blood pressure artificially and put less strain on the system. Unfortunately, these drugs have many ill side effects and, in addition, treat only the symptoms of high blood pressure, not the underlying causes.

A number of studies have been conducted on the effects of the TM technique on patients with high blood pressure. In one of the early studies, in which I was involved, blood pressure was recorded 1,119 times in 22 borderline hypertensive patients before and after beginning the TM program, over 4-63 weeks. A significant average decrease of 9 mm Hg systolic and 6 mm Hg diastolic was found. This was independent of anti-hypertensive medication. The study did not include controls; however, the blood pressure of those subjects who stopped practicing TM returned to premeditation levels.

In another study, Blackwell and coworkers (1975) investigated both clinical and home-recorded blood pressures in seven meditating subjects, reporting decreases of up to 13 mm Hg for systolic and 7.3 mm Hg for diastolic pressures. Blood pressure recordings in all these studies were taken randomly throughout the day, demonstrating the significant carry-over effect from the meditation period. Simon and coworkers (1974) at the University of Chicago followed a group of borderline hypertensives for a mean of eight months and found a significant decrease of 15 mm Hg systolic and 9 mm Hg diastolic. In one of several other studies Agarwal and Kharbanda (1979) found similar improvements in mild and moderate hypertensive patients in India.

The second cardiovascular risk factor, high cholesterol levels, has been known for a number of years to be correlated with the occurrence of arteriosclerosis and coronary heart disease. More recently, attention has turned to the group of compounds known as the lipoproteins. During stress there may be a discrepancy between fat mobilization and utilization, which in turn may result in an increased synthesis and secretion of very low-density lipoproteins by the liver. High levels of very low-density lipoproteins have been shown to increase damage to the wall of the blood vessels and produce other symptoms associated with the development of arteriosclerosis. Some researchers suggest that this damage to the smooth inner lining of the walls of the artery starts the mechanisms that

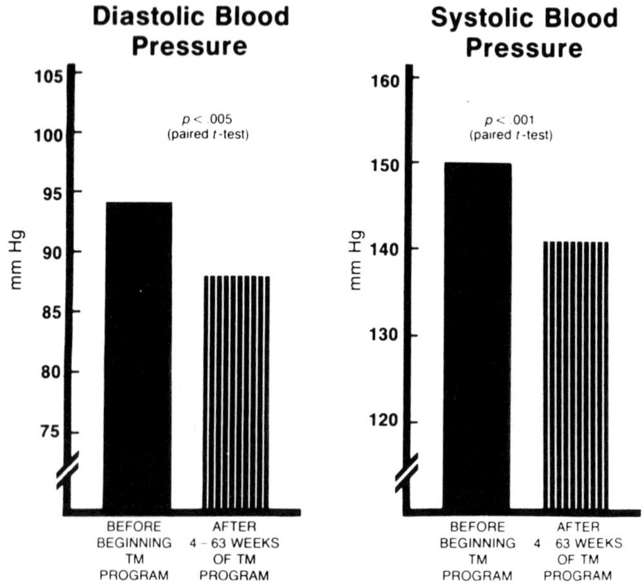

Finding: Systolic and diastolic arterial blood pressures were recorded 1,119 times in 22 hypertensive patients before and after they began the Transcendental Meditation program. The decreases in blood pressure after patients began practicing the Transcendental Meditation technique were statistically significant.

First Reference: "Decreased Blood Pressure in Hypertensive Subjects Who Practiced Meditation," Supplement II to *Circulation* 45 and 46 (U.S.A.: 1972).

Second Reference: Barry Blackwell, Irwin Hanenson, Saul Bloomfield, Herbert Magenheim, Sanford Nidich, and Peter Gartside, "Effects of Transcendental Meditation on Blood Pressure: A Controlled Pilot Experiment," *Journal of Psychosomatic Medicine* 37, (U.S.A.: 1975): 86.

EFFECT ON BLOOD PRESSURE

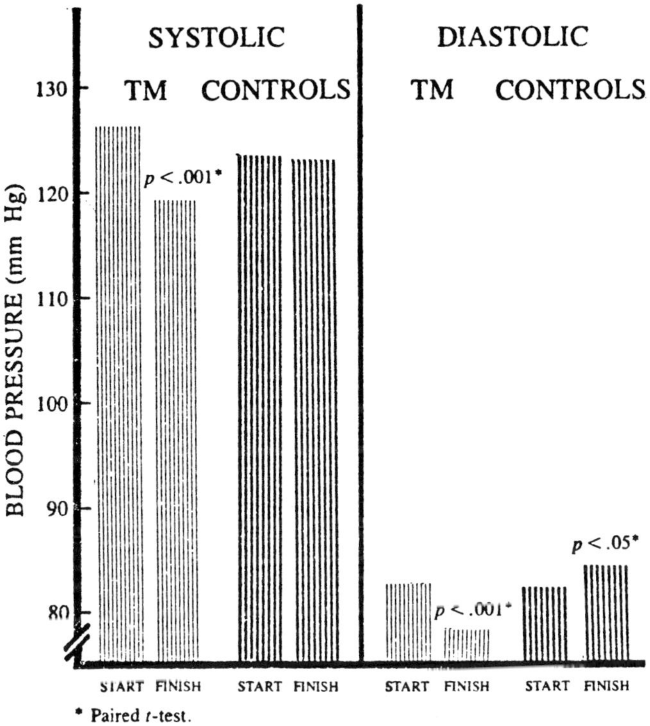

CHANGES IN BLOOD PRESSURE DURING A PERIOD OF 10–14 MONTHS FOR 23 SUBJECTS WHO LEARNED THE TRANSCENDENTAL MEDITATION TECHNIQUE AT THE BEGINNING OF THE PERIOD AND 21 CONTROLS WITH NO TREATMENT. The TM group showed a significant decrease of both diastolic and systolic blood pressures ($p < .001$) whereas the diastolic blood pressure of the control group was found to increase ($p < .05$).

COOPER, M. J., and AYGEN, M. M. Effect of Transcendental Meditation on serum cholesterol and blood pressure. Findings previously published in *Harefuah,* the Journal of the Israel Medical Association, 95(1): 1–2, 1978.

actually lead to arteriosclerosis. This break in the smoothness of the wall may create a stress point at which further damage may result.

The effects of the TM technique on high cholesterol have been investigated by Cooper and Aygen (1979). They studied 23 hypercholesterolemic patients and controls matched for age, sex, diet, weight, and initial cholesterol values. Twelve of the subjects participated in the TM program for 11 months; their serum cholesterol significantly dropped, while the cholesterol level of non-meditating controls did not change. The observed decreases in subjects' cholesterol levels were significant compared to their own base line and to controls. In another study on normal subjects, Cooper and Aygen (1978) also showed a significant decrease in cholesterol levels and blood pressure over several months.

Studies reporting beneficial effects of the TM program on two other cardiovascular risk factors, smoking and obesity, have also been conducted (Shafii et al., 1976; Bauhofer, 1983). For example in a retrospective study of 1080 individuals practicing the TM technique reported a significant decrease in cigarette smoking. In another retrospective study, Weldon and Aron (1974) surveyed a group of college students and older TM participants and found that overweight subjects tended to loose weight and underweight subjects tended to gain weight.

The last and most recent risk factor to be identified concerns a personality or behavior type known as the type A coronary-prone behavior pattern. First identified by Rosenman and Friedman (1974), this pattern is characterized as a particular action-emotion complex that can best be described as a compulsive inappropriate response to stress. The type A individual can be distinguished from someone suffering from anxiety or fear: he may struggle, but he rarely despairs of losing. Instead, he shows certain enhanced traits—aggressiveness, ambitiousness,

Decreased Serum Cholesterol
(in Hypercholesterolemic Subjects)

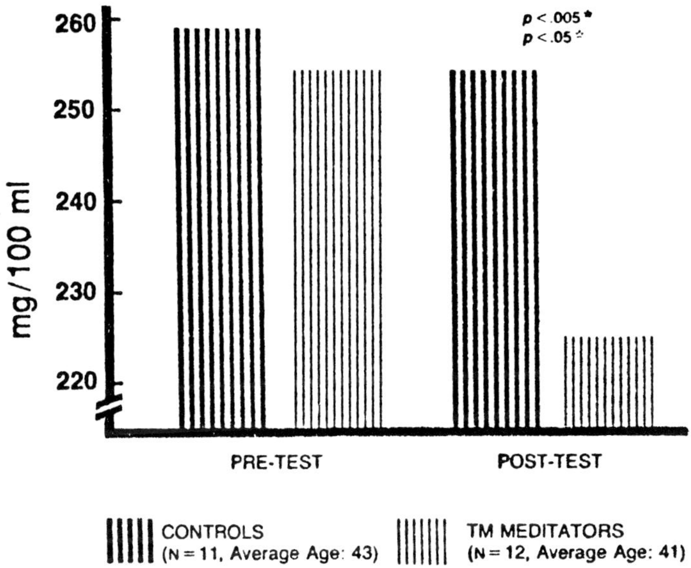

* Compared to pretrial (by paired *t*-test).
☆ Compared to change in controls (by unpaired *t*-test).

Research has shown that Transcendental Medita tion reduces serum cholesterol, a major risk factor for heart disease. In this study, serum cholesterol was reduced in a group of subjects with hypercholesterolemia who practiced Transcendental Meditation over an 11-month period, while no change was observed in a control group. Transcendental Meditation has also been found to reduce serum cholesterol to more ideal levels in subjects within the normal range

COOPER, M. J., and AYGEN, M. M. Transcendental Meditation in the management of hypercholesterolemia. *Journal of Human Stress* 5(4): 24–27, 1979.

SERUM CHOLESTEROL

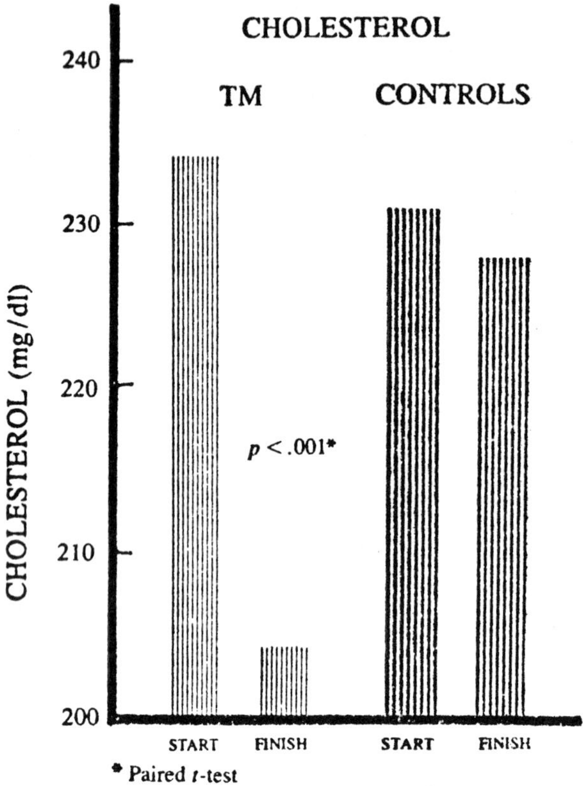

CHANGES IN SERUM CHOLESTEROL LEVELS DURING A PERIOD OF 10–14 MONTHS FOR 23 SUBJECTS WHO LEARNED THE TRANSCENDENTAL MEDITATION TECHNIQUE AT THE BEGINNING OF THE PERIOD AND 21 CONTROLS WITH NO TREATMENT. The TM group showed a significant decrease of serum cholesterol concentration ($p < .001$). No significant change was seen in the no-treatment control group.

COOPER, M. J., and AYGEN, M. M. Effect of Transcendental Meditation on serum cholesterol and blood pressure. Findings previously published in *Harefuah,* the Journal of the Israel Medical Association, 95(1): 1–2, 1978.

Use of Alcohol and Cigarettes

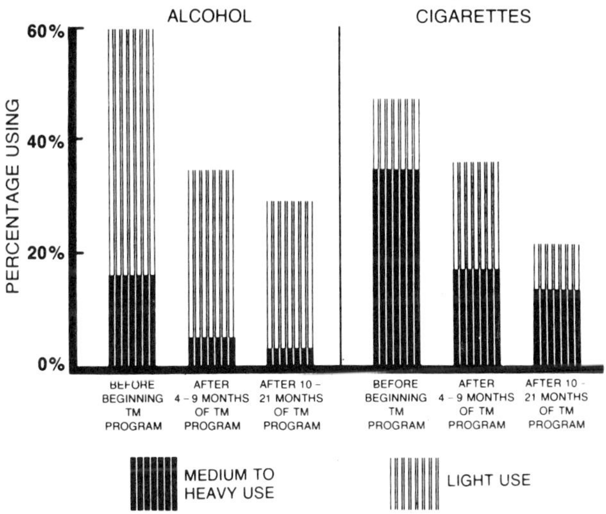

Finding: A retrospective study of 1,862 subjects who had practiced the Transcendental Meditation technique an average of 20 months showed a significant reduction in the reported use of alcohol and cigarettes.

Reference: "Decreased Drug Abuse with Transcendental Meditation: A Study of 1,862 Subjects," *Drug Abuse: Proceedings of the International Conference*, ed. Chris J. D. Zarafonetis (Philadelphia, Pennsylvania, U.S.A.: Lea and Febiger, 1972): 369–376 and *Congressional Record*, Serial No. 92–1 (Washington, D.C., U.S.A.: Government Printing Office, 1971).

competitiveness—which, when activated by environmental interactions, may result in hostility or chronic impatience. In contrast to the type A coronary-prone individual is the type B person, in whom type A characteristics are relatively absent.

A number of studies over the past 20 years have found considerable evidence supporting the hypothesis that type A behavior is highly correlated with the occurence and prevalence of coronary heart disease (The Review Panel on Coronary-Prone Behavior and Coronary Heart Disease, 1981). The cornerstone of this evidence came from the Western Collaborative Group Study (Rosenman et al., 1975), which followed 3154 middle-aged men over nearly nine years. Approximately 50 percent of this group exhibited type A characteristics and demonstrated almost twice the rate of new coronary heart disease as did the type B individuals. The risk of recurrent heart disease was one-and-one-half times greater in the type A's. Several retrospective studies have consistently shown that coronary heart disease patients scored higher on the type A behavior identifying test than patients who were hospitalized with other ailments and non-hospital controls.

Research on the relationship of type A coronary-prone behavior to traditional coronary heart disease risk factors such as cholesterol level, blood pressure, and cigarette smoking indicates that type A behavior is a separate risk factor largely independent of other standard risk factors. Other risk factors, however, appear to be more prevalent in type A persons. For example, several studies have shown that type A people apparently have higher cholesterol levels than type B people.

A number of researchers are now attempting to identify the physiological pattern of responses that distinguishes type A from B individuals (Ward et al., 1986). The findings so far indicate that type A individuals show an accentuated biochemical and physiological response, especially in stressful situations. For example, there are a number of biochemicals and hormones, such as adrenalin and cortisol, that naturally increase in response to stressful situations. It has been shown that type A's exhibit much higher levels of these compounds than type B's, especially in stressful situations. Further, a number of investigators have attempted to determine the difference in cardiovascular responses between type A and type B individuals.

In general, the studies have shown that when type A individuals are placed in a challenging situation, i.e., when their competitive nature is provoked, they exhibit a significantly higher increase in blood pressure than the type B individuals. These and other studies have led researchers to speculate on the specific mechanisms by which the type A person may be more susceptible to coronary heart disease. There may be a hemodynamic factor—excessive and prolonged fluctuations in blood pressure—that might strain and possibly damage the smooth linings of the coronary artery walls. Chemical factors, such as elevation in very low-density lipoproteins or cholesterol levels, might also lead to damage and eventually arteriosclerosis.

Several studies have indeed related the type A coronary-prone behavior pattern with the severity of coronary arteriosclerosis. Researchers are also finding clues in the activation of the sympathetic nervous system in type A subjects, and the resulting increase in adrenalin and similar components. Arteriosclieriosis partially blocks the blood vessels and thus restricts the flow of oxygen to the heart muscles. An increase in adrenalin consequently may place too great a demand on the oxygen consumption capability of these muscles, and thus may further enhance susceptibility to coronary heart disease. Activation of the sympathetic nervous system may also increase the likelihood of an abnormal rhythm within the heart, what can also lead to a heart attack. Thus, many physiological factors associated with type A behavior may in fact contribute to susceptibility to coronary heart disease.

If type A individuals are, indeed, characterized by an excessive physiological response to environmental demands, then one approach to decreasing their susceptibility to coronary heart disease would be to change this response pattern. Extensive research on the Maharishi Technology of the Unified Field indicates, as we have mentioned, that during the TM technique participants obtain a state of deep rest that is in many ways physiologically opposite to the response seen in stress tests. In addition, TM meditators have been shown to have an improved and more rapid adaptation to stress.

As we have also discussed, research by Orme-Johnson, Goleman and Schwartz, and by Daniels, using a variety of physiological measures, showed that meditators recovered significantly faster than non-meditators following stressful stimuli. These studies suggest that the TM technique would have a very important effect on type A behavior and susceptibility to heart disease. Altering the individual's physiology so that he is more resistant to the deleterious effects of stress might make him better able to cope and succeed, even in the fast pace of life today.

Two further studies on the effects of TM on cardiovascular disease investigated patients with angina pectoris and arrhythmia. Zamarra and coworkers (1975) at the State University of New York at Buffalo studied the effects of the TM technique on angiographically documented angina pectoris. Patients with classic stable angina pectoris for six months or more were pretested two or three times on a variety of exercise tolerance tests. One group of 10 patients then began the TM technique, and the others acted as a control group and were told they would later receive instruction in TM. All subjects were retested after eight months.

The 10 patients in the TM group showed a 14.5 percent increase in exercise tolerance, an 11 percent increase in maximum work load, a 16 percent increase in delay of ST segment depression, and a 7.25 percent decrease in double product (proportional to heart rate x systolic blood pressure). There were no significant changes in six control patients. The authors conclude that the "regular practice of the TM technique is associated with modification of the hemodynamic response to stress and an increase in exercise capacity in patients with stable angina pectoris." Lowns and coworkers (1976), in a short communication, also reported that the TM program was useful as part of the treatment of recurring ventricular arryhthmia.

OTHER CLINICAL STUDIES

A number of studies have also been conducted on other clinical and health-related conditions. The effects of the TM technique on bronchial asthma were evaluated in a six-month random assignment crossover study by Wilson and coworkers (1975) at the University of California at Irvine. Twenty-one stable asthmatic patients were randomly divided into two groups and were evaluated by a variety of pulmonary function tests, a symptom severity questionnaire and diary, as well as physical evaluations.

Group A began and practiced the TM technique twice a day while group B began reading the book *The Science of Being and Art of Living* by Maharishi Mahesh Yogi, which describes the theory and benefits but not the detailed instruction of the TM technique. After three months the groups crossed over, that is, Group A was asked to stop meditating and start reading and Group B was instructed to stop reading and to start the practice of the TM technique. After three months of practicing the TM technique Group A showed a significant improvement in forced expiratory volume, peak expiratory flow volume, and airway resistance. When they ceased meditating for the last three months of the study their respiratory functions declined again towards their initial levels.

In Group B, after the first three months of reading there was no significant change in any respiratory functions after the second three months, during which they practiced the TM technique, their airway resistance significantly improved. Six and twelve-month evaluations assessed the severity of asthma symptoms by questionnaires, a daily diary, and physicians' evaluations. In the majority of subjects, the TM program had markedly improved the asthmatic condition. The authors concluded that the TM program may be a useful adjunct in the treatment of asthma (Honsberger and Wilson, 1973a, 1973b; Wilson et al., 1975).

In another study which supports these findings, Corey (1973) reported an increase in specific airway conductance and forced expiration volume in TM subjects as compared with controls, thus indicating general improvement in respiratory functions even in normal subjects.

The TM program has been found useful in the treatment of insomnia. Miskiman (1972b) studied a group of medically referred chronic insomniacs who reported that time taken to fall asleep ranged from 78 minutes prior to learning the TM technique to 15 minutes after learning it. These changes were still intact at a one year follow-up. Fuson (1976) replicated the results of this study. Miskiman (1972a) also reported faster recovery from sleep deprivation in TM subjects versus controls.

The TM program has been shown to be beneficial in the treatment of a number of major and minor psychiatric disorders (Rigby, 1977; Suurkula, 1977; Egerman, 1981; Wood, 1981; Brooks and Scarana, 1985). For example, in an outpatient setting, Dick and Ragland (1973) found that clients of a university counseling center who were randomly assigned to learn the TM program showed significantly greater positive changes on a variety of personality

Change in Bronchial Asthma

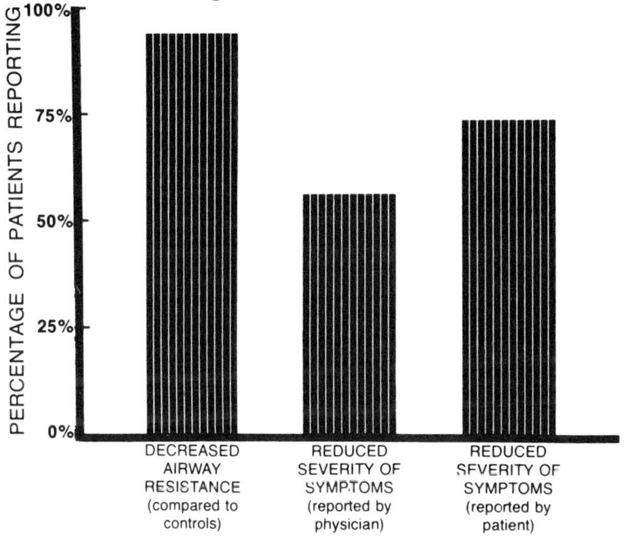

Finding: After beginning the practice of the Transcendental Meditation technique 94 percent of a group of asthmatic patients showed improvement as determined by the physiological measurement of airway resistance. Fifty-five percent of the asthmatic patients showed improvement as reported by their personal physicians, and 74 percent showed improvement as reported by the patients themselves.

First Reference: Ron Honsberger and Archie F. Wilson, "The Effect of Transcendental Meditation upon Bronchial Asthma," *Clinical Research* 22, no. 2 (U.S.A.: 1973)

Second Reference: Ron Honsberger and Archie F. Wilson, "Transcendental Meditation in Treating Asthma," *Respiratory Therapy: The Journal of Inhalation Technology* 3, no. 6 (U.S.A.: 1973): 79–80.

Third Reference: Archie F. Wilson, Ron Honsberger, J.T. Chiu, and H.S. Novey, "Transcendental Meditation and Asthma," *Respiration* 32 (U.S.A.: 1973): 74–80.

Fourth Reference: Paul W. Corey, "Airway Conductance and Oxygen Consumption Changes Associated with Practice of the Transcendental Meditation Technique," (University of Colorado Medical Center, Denver, Colorado, U.S.A.).

Change in Time of Sleep Onset

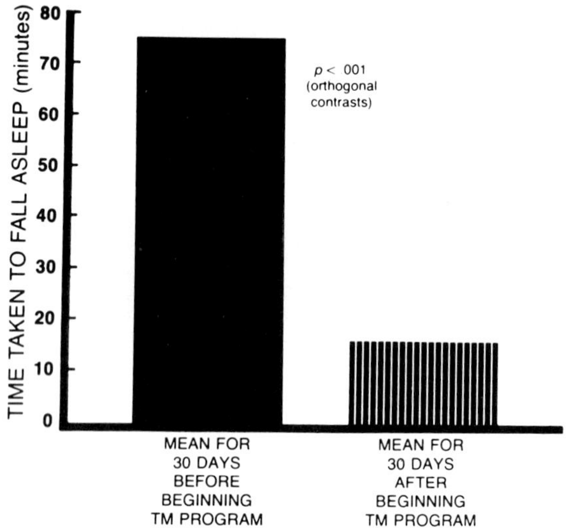

Finding: The Transcendental Meditation technique significantly reduced the time taken for insomniacs to fall asleep. As a therapy against insomnia, the Transcendental Meditation program was reported to be simple to administer, immediately effective, stable over time, and without unfavorable side effects.

First Reference: Donald E. Miskiman, "The Treatment of Insomnia by the Transcendental Meditation Technique," (University of Alberta, Edmonton, Alberta, Canada).

scales over eight weeks compared with controls practicing an eyes-closed rest, placebo-type practice.

In a comprehensive study on TM and childbirth involving over 100 mothers it was shown that the TM subjects had better health during pregnancy and childbirth, shorter duration of labor, and a lower frequency of operative intervention during labor (Heidelberg, 1979).

In addition to these studies, a survey of health problems has reported improvements in a number of stress-related conditions such as ulcers, physical tension, and headaches, as well as in behavioural problems such as excessive alcohol consumption (Wallace, 1970a; Shafii et al., 1975; Lovell-Smith, 1982). Further, McIntyre and coworkers (1975) and Allen (1979) also found that stuttering subjects reported an improvement in their speech pathology. That the TM technique causes so many different improvements strongly supports the idea that it removes deep-rooted abnormalities and allows for a normalization of physiological functioning.

Finally, perhaps the most important health-related study to date examined health insurance statistics for TM meditators. This study showed that the rate at which meditators use medical and surgical health care services is approximately one-half that of other insurance users. The study controlled for other factors that might affect health care use, such as cost-sharing, age, gender, geographic distribution, and profession and found that they could not account for the meditators' lower rates. In addition, the TM meditators showed a much lower rate of increase in health care utilization as a function of age than did comparison groups (Orme-Johnson, 1986).

DRUG ABUSE

A number of studies have reported the beneficial effects of the TM program in helping individuals stop

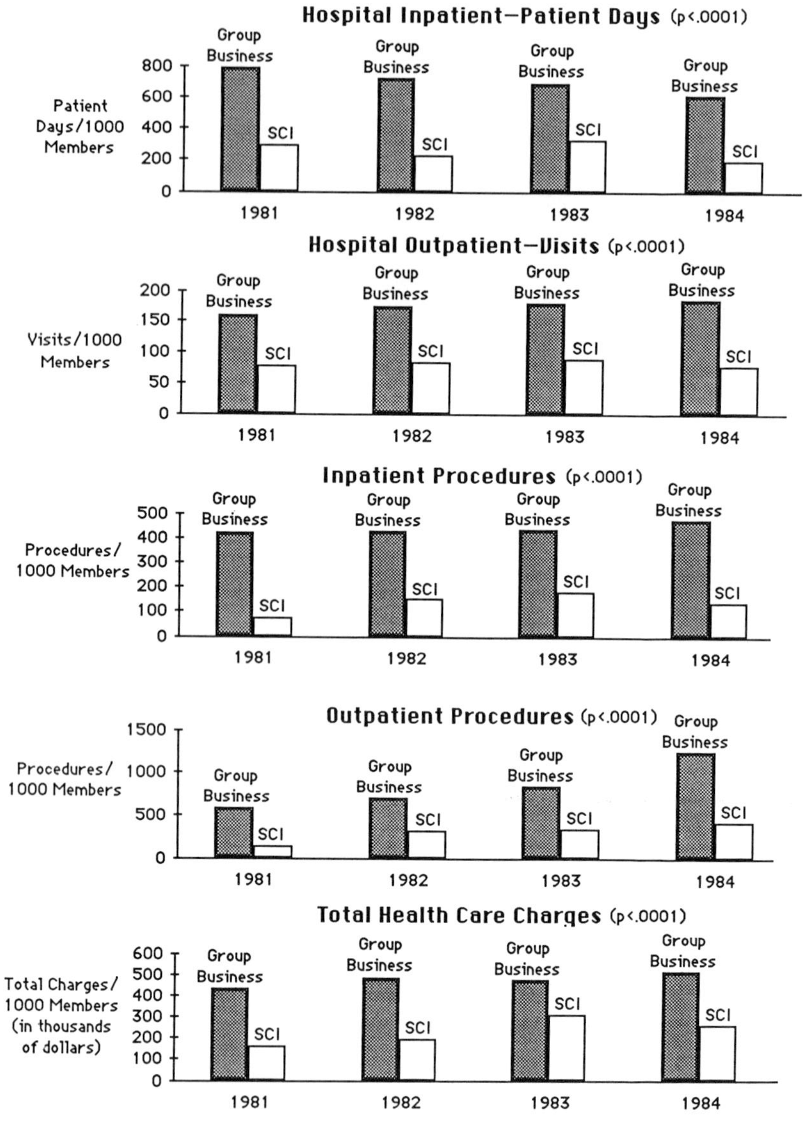

Figure 1. Health Care Utilization Statistics Comparing the SCI Medical Insurance Group (Regular Practitioners of the Transcendental Meditation Program) and Group Business (the Mean of All Iowa Insurance Groups with the Same Carrier).

SCI utilization rates in all categories were consistently lower than those for Group Business over all four years of the study.

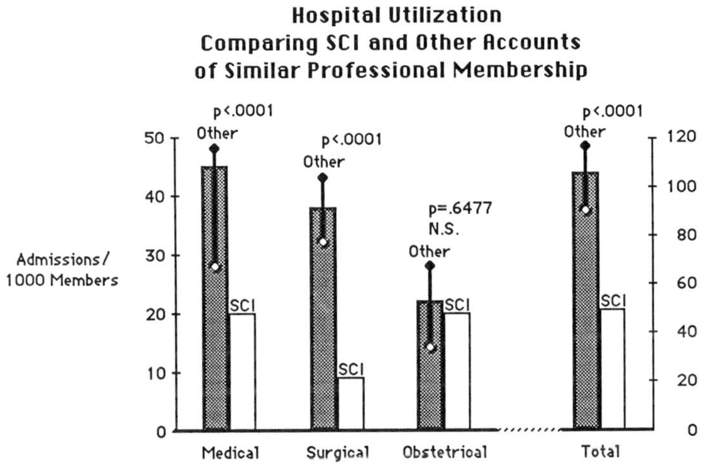

Figure 5. Hospital Utilization of the SCI Health Insurance Group Compared to Other Accounts of Similar Professional Membership.

This figure shows the comparison of the SCI group with Other Accounts of similar professional membership selected by the actuarial service of the insurance carrier. Hospital admissions for SCI were 55.6% less than the average of the Other Accounts for medical, 76.3% less for surgical, similar for obstetrical, and 53.3% less for total admissions. The error bars show the range of the Other Accounts, indicating that the SCI Group had lower admission rates than even the lowest of the comparison group on medical, surgical, and total admissions. P values are for comparison of SCI with the average of Other Accounts

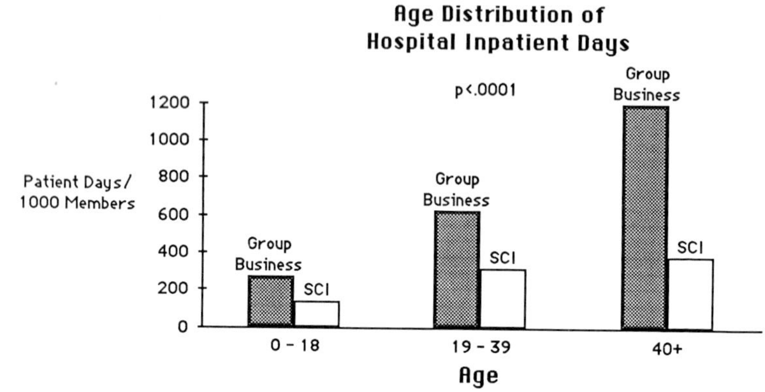

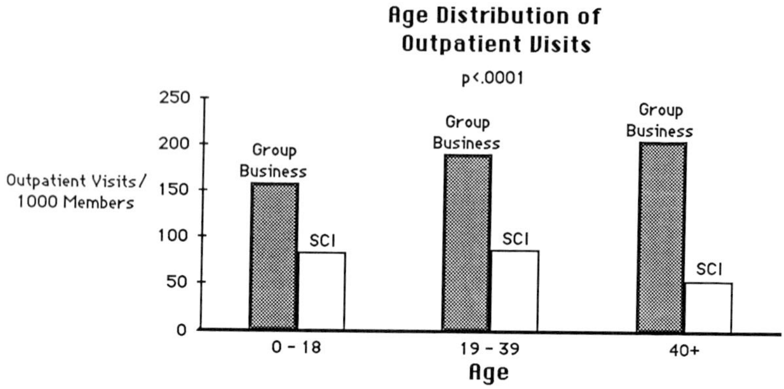

Figure 4. Health Care Utilization by Age Group for the SCI Medical Insurance Group Compared to Group Business. The rates of hospital inpatient days and outpatient visits typically increase with age, as can be seen above in the statistics for Group Business. The SCI group, however, not only showed much lower

abusing drugs. In an initial retrospective study conducted by myself and coworkers, questionnaires were administered to 1,950 TM subjects, with 1,862 completed responses. The population consisted primarily of college students who had been practicing the technique for an average of 20 months.

After they began the TM technique, the number of drug abusers significantly decreased in all categories studied: marijuana, hallucinogens, narcotics, amphetamines, and barbiturates. The longer the subjects practiced TM, the more their abuse declined, until after 21 months of meditation most subjects had stopped abusing drugs.

Shafii and coworkers (1974) administered a questionnaire to compare marijuana use in TM participants and controls. The researchers obtained the controls by asking each subject for the names of two other students who were similar in age, sex and educational level, but were not practicing the TM technique. Almost half of those who had practiced the TM technique for a period of one to three months decreased or stopped their use of marijuana. Over the same period less than one-sixth of the controls decreased or stopped their usage.

The longer a person had practiced meditation, the higher was the probability that marijuana use would be discontinued. Of those who had practiced the TM technique for more than two years, 92 percent had significantly decreased their use of marijuana, while 77 percent had totally stopped. Again, less than one-sixth of the control group showed similar results. In another study, Shafii and coworkers (1975), using a similar self-report design with matched controls, found a marked reduction in alcohol consumption in TM subjects as compared with little or no change in the controls.

Monahan (1977) used a similar type of design in which TM subjects asked a non-meditating friend to complete

Drug Use

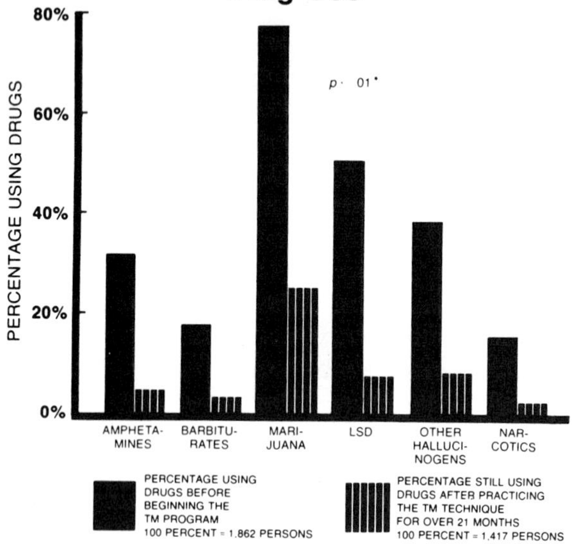

*t-test for marijuana usage, second reference.

Finding: A retrospective study of 1,862 subjects who practiced the Transcendental Meditation technique an average of 20 months showed decreases in the reported use of non-prescribed drugs.

Among the 852 subjects who had practiced the TM technique over 22 months, only 12.2% used marijuana, 3% used LSD, 4% used other hallucinogens, 1.2% used narcotics, 1.2% used amphetamines, and 1% used barbiturates. Nearly all these subjects who continued to use drugs reported using them only very rarely.

First Reference: "Decreased Drug Abuse with Transcendental Meditation: A Study of 1,862 Subjects," *Drug Abuse: Proceedings of the International Conference*, ed., Chris J. D. Zarafonetis (Philadelphia, Pennsylvania, U.S.A.: Lea and Febiger, 1972): 369–376 and *Congressional Record*, Serial No. 92–1 (Washington, D.C., U.S.A.: Government Printing Office, 1971).

and return a questionnaire. The results of the study showed significant decreases in drug use for TM participants across all categories of substances with no consistent change for controls.

Lazar and coworkers (1972) studied self-reported marijuana, alcohol, and tobacco use in college students and adults just prior to TM instruction and at either four, eight, or twelve-week intervals after instruction. Findings indicated sharp declines across the three categories as well as significant reductions in anxiety. Decreased anxiety was found to be correlated with decreased use of drugs.

Katz (1974) conducted a prospective study of high school and college students. A drug history questionnaire was distributed to 268 subjects who had decided to learn the TM technique. The subjects' anonymity was maintained to insure frank answers. Follow-up questionnaires were later mailed to all subjects after they learned TM. As in previous studies, controls were obtained by subjects asking non-meditating friends to complete and return questionnaires over a similar time period. In the pretest, 198 controls completed the questionnaire. One-hundred-and-fifty TM subjects and 110 controls returned at least one follow-up questionnaire. Subjects practicing the TM technique significantly decreased their use of marijuana, hashish, wine, beer and "hard liquor" in comparison with the control group.

Several studies have evaluated the beneficial effects of the TM program in drug abuse treatment programs (Brautigam, 1972; Schenkluhn and Geisler, 1974). For example, Schenkluhn and Geisler studied 76 patients at a drug rehabilitation center in Germany and found a significant drop in the use of all drugs, including amphetamines, barbiturates, and opiates after 12 months of TM practice.

A number of studies have also reported improvements in drug abuse in prisoners (Orme-Johnson, 1981; Ballou,

1973a, 1973b; Ramirez, 1976; Ferguson, 1978).

In an excellent review of the TM program and the treatment of drug abuse, Siegel (1978) concludes that:

> ...the positive findings among TM participants in these studies would appear to be of sufficient magnitude to seriously consider the implementation of the TM program as a major treatment modality in both community and institutional drug treatment programs. The TM program is simple to administer, inexpensive, easily integrated into existing programs, and does not disrupt program or institutional routines. Results of work to date indicate the effectiveness of the TM program in reducing many problems in the drug abuser's life, while simultaneously improving the capacity to live a more fully developed, useful, and enjoyable life.

OTHER AYURVEDIC APPROACHES TO PERFECT HEALTH

The research on the Maharishi Technology of the Unified Field demonstrates the effectiveness of the Ayurvedic approach to perfect health on the level of consciousness. Ayurveda also includes a number of other approaches to health on level of physiology, behavior and environment.

One of the most important concepts in Ayurveda, according to the classic text of Ayurveda, the Charaka Samhita (Sharma, 1981), is an understanding of different psycho-physiological constitutions. Each individual can be broadly categorized into one of seven different body types (some authors refer to ten types). These types are based on the balance or proportion of three fundamental physiological principles, referred to as the three "doshas"— vata, pitta and kapha. The three doshas may be found in different proportions in different individuals, yet proper health in any individual is dependent upon maintaining them in an appropriate balance. Aggravation or imbalance leads to disorder and disease (Dubey and Singh, 1970).

The only equivalent concept in modern medicine is type A and B coronary-prone behavior. There have been a number of attempts throughout the history of Western medicine to develop more precise schemes of psycho-somatic typing, the best known being Sheldon's typing of ectomorph, mesomorph and endomorph, but none of these approaches have been practically applied. Only the traditional systems of medicine, such as those found in India and China, have maintained an emphasis on different body types.

Body typing is a fundamental part of Ayurveda, for it plays a significant role in many prevention measures as well as in the diagnosis and treatment of disease. Unfortunately, terminology utilized to describe the different types is quite general, and very little scientific research has been conducted in order to understand these types in terms of objective physiological criteria. Let us describe the three doshic principles underlying these psycho-physiological body types and then briefly review the research.

PSYCHO-PHYSIOLOGICAL BODY TYPES

The principle of vata is associated with movements in the body. The elements of air and space underlie its tendencies. Vata is involved in such functions as respiration, excretion, and neural control of sensory and motor functions. Imbalance in vata causes many mental and physical disorders including anxiety, fear, pain, backache, constipation, sciatica and menstrual problems.

The principle of pitta is involved with metabolic transformation and bioenergetics. The elements of fire and, to some extent, water are at the basis of pitta tendencies. Pitta is associated with such functions as metabolism, digestion and thermo-regulation. Disorders considered to be due to pitta imbalance include anger, hostility, peptic ulcers and hyperthyroidism.

Kapha is concerned with the structural basis of the physiology. Its characteristics of solidity and inertia come from the predominance of the earth and water elements in its nature. Structures such as membranes and connective tissue, which underlie the connectedness and stability of the different parts of the body, are associated with kapha. Bronchial disorders, asthma, disorders related to inflammation of the joints, and obesity occur as a result of kapha imbalances.

Susceptibility to different diseases, as well as the course of development of each disease, will vary depending upon the predominance of these three doshas in the individual. Thus, establishment of the individual's body type is crucial for diagnosis as well as treatment.

Considering the potential importance of psychosomatic typing, it is remarkable that so little scientific research has been conducted in this area. Perhaps this is because of the great variability among individuals and the lack of fundamental understanding of the physiological basis of such a typing scheme. A considerable amount of research has been conducted on type A and B coronary-prone behavior, but researchers are cautious about its accuracy and therefore its effectiveness as a reliable diagnostic tool.

I believe that the understanding provided by Ayurveda could give the proper theoretical basis for a thorough scientific investigation in this area. Preliminary evidence in our laboratories has shown some interesting results. For example, in one study subjects were assessed according to the traditional procedures of Ayurveda and also by the Jenkins Activity Index for Type A and B behavior. Significant correlations were found between individuals with predominantly pitta constitution and the type A coronary-prone behavior pattern, and those with kapha constitution and the type B behavior pattern. In addition, the kapha type individuals had significantly higher cholesterol and triglyceride levels than the vata type individuals. Further, significant differences were also found in measurements of pulse rate, white blood cell counts, and EEG frequencies. The vata type had the highest pulse rate, while the kapha type had the highest white blood cell count and also the slowest frequency and highest amplitude of alpha wave activity (Schneider et al., 1985a).

One of the most interesting investigations concerning the physiological basis of Ayurvedic body types was conducted in India at the University of Benares. In these studies an attempt was made to correlate the three main Ayurvedic body types with blood levels of three principle neurotransmitters. The findings over several different studies indicate that the vata type has significantly higher levels of acetylcholine, the pitta higher levels of norepinephrine, and the kapha higher levels of histamines (Udupa et al., 1975; Singh et al., 1984).

Further research in this area should provide a fuller physiological understanding of the Ayurvedic psychosomatic typing system and thus help introduce it more widely to the West. On the basis of this typing system, Ayurveda provides an elaborate nutritional and dietary program, specifically designed for each type. Further, it gives specific advice and procedures in regard to behavior, lifestyle and physical fitness. Extensive knowledge exists concerning daily and seasonal biological

Therapeutic Strategies of Ayurveda

Ayurveda advocates four complementary approaches that contain multiple therapeutic and preventive strategies.

---

1 **CONSCIOUSNESS**—utilizing the Transcendental Meditation and TM-Sidhi program,* the subjective aspect of the Maharishi Technology of the Unified Field. This technology allows the mind to identify itself with the unified field of all the laws of nature, thereby enlivening the quality of perfect balance in all aspects of mind, body, behavior, and environment.

---

2 **PHYSIOLOGY**—applying extensive dietetic measures, sophisticated purification procedures, as well as herbal and mineral preparations for the treatment of specific disorders, the enhancement of immune system functioning, and the promotion of longevity.

---

3 **BEHAVIOR**—providing comprehensive guidelines for personal hygiene, exercise, and health-promoting conduct as well as daily and seasonal routines to restore integrity to fundamental biological rhythms.

---

4 **ENVIRONMENT**—eliminating stress and increasing balance in society as a whole through collective health measures, the most effective of which is the group performance of the Maharishi Technology of the Unified Field.

RELATION OF TYPE A BEHAVIOR TO
AYURVEDIC PSYCHOSOMATIC TYPE

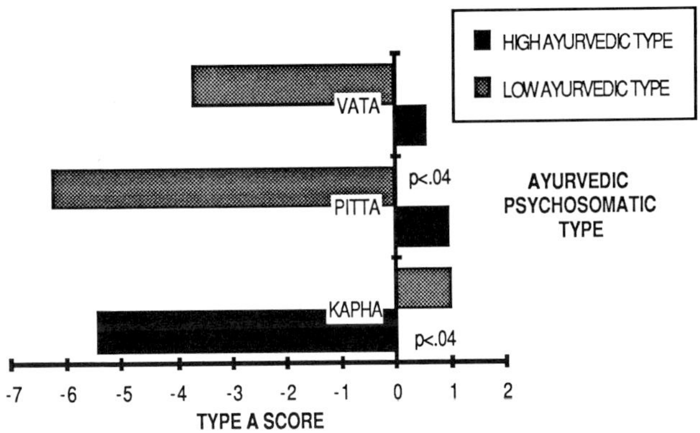

Figure 3. **Psychological Correlates of Ayurvedic Psychosomatic Types: Relation to Type A Coronary Prone Behavior.** Individuals with extreme Pitta constitutions were rated as more Type A by the Jenkins Activity Survey [JAS] than Vata or Kapha types. In contrast, extreme Kapha individuals were rated as strongly Type B.

Reference*: R.H. Schneider, D. Orme-Johnson, J. Kesterson, R.K. Wallace: Physiological and Psychological Correlates of Ayurvedic Psychosomatic Types. Paper presented at the Eighth World Congress of the International College of Psychosomastic Medicine, Chicago, IL, 7 Sept. l985.*

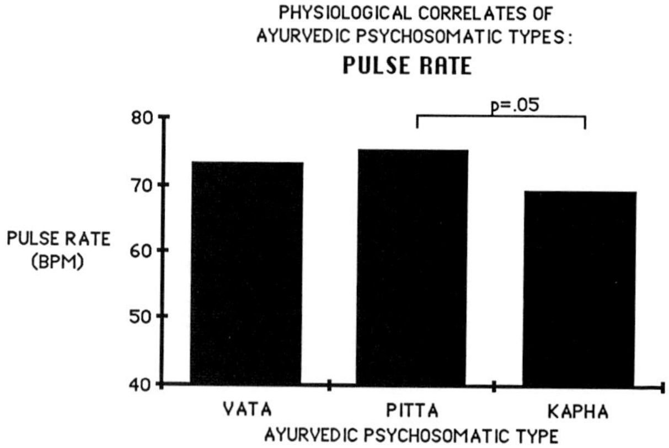

Figure 4. **Physiological Correlates of Ayurvedic Psychosomatic Types: PULSE RATE.** Subjects who were diagnosed by Ayurvedic technique as extreme Kapha were found to have significantly slower resting pulse rates than extreme Pitta types. Vata individuals also tended to faster pulse rates than Kapha types.

Reference: *R.H. Schneider, D. Orme-Johnson, J. Kesterson, R.K. Wallace: Physiological and Psychological Correlates of Ayurvedic Psychosomatic Types. Paper presented at the Eighth World Congress of the International College of Psychosomastic Medicine, Chicago, IL, 7 Sept. 1985.*

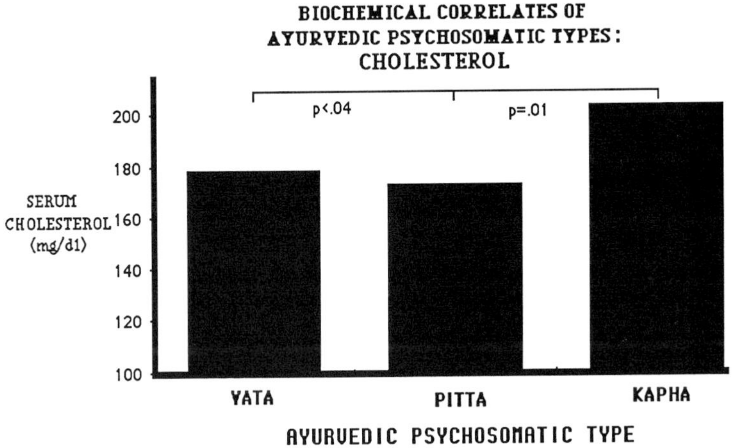

Figure 5. **Biochemical Correlates of Ayurvedic Psychosomatic Types: CHOLESTEROL.** All within the normal range, subjects rated as predominanty Kapha had higher levels of cholesterol in the blood than predominantly Vata or Pitta types.

Reference: *R.H. Schneider, D. Orme-Johnson, J. Kesterson, R.K. Wallace: Physiological and Psychological Correlates of Ayurvedic Psychosomatic Types. Paper presented at the Eighth World Congress of the International College of Psychosomastic Medicine, Chicago, IL, 7 Sept. l985.*

rhythms with regard to individual physiological type. On this basis Ayurveda prescribes certain daily and seasonal routines for better health.

NATURAL MEDICINE

Over the many thousands of years, the Ayurvedic tradition has reserved a very large body of knowledge concerning the medicinal value of preparations made from of specific plants, fruits, metals, and minerals. The procedures for these preparations are often elaborate and highly sophisticated. The recommended use of these herbal and mineral preparations is again based on a thorough understanding of the individual body type as well as medical history.

Numerous biochemical and pharmacological studies have been conducted on the traditional herbs of Ayurveda as well as those of other traditional systems. These studies, however, often are reductionist in their approach, with the primary aim of isolating an active ingredient which can then be synthesized and marketed as a modern drug. Ayurveda specifically emphasizes the synergistic nature of its medicinal preparations, which may contain a combination of herbs with several active ingredients that, when properly mixed, eliminate the harmful side effects so common in modern medicine. There is a need for a more complete understanding of these preparations with reference to other factors Ayurveda regards as important, such as diet, season, and mental state.

Ayurveda places the real value of the herbs and minerals in terms of the overall effect they produce, not only on a gross physical level, but on the holistic value of consciousness. These preparations in effect stimulate, on the level of both consciousness and physiology, the level of intelligence or natural law corresponding to the intelligence in the herb. The ultimate effect is to restore balance to the whole system rather than treat a specific symptom.

One group of herbal fruit and mineral preparations, referred to as rasayanas, is specially used as a preventive measure to retard the aging process. A recent study, conducted at MIT, has shown some highly interesting results from administering one of the rasayanas to animals undergoing rapid aging. Two groups of animals were given a choline deficient diet which accelerates the aging process and intentionally causes necrosis of kidney tissue. The experimental group was given a specific Ayurvedic rasayana which, when chemically analyzed, was shown not to contain choline. This group showed no signs of kidney tissue necrosis while the control mice showed extensive damage. Further studies on the effects of these different rasayanas have also shown promising results in the prevention of cancer in certain animals (Abou Nader, 1986).

Ayurveda specifies that these rasayanas are most effective following a series of purification procedures known as panchakarma. Panchakarma involves therapies such as herbalized massage, steam treatments, and a variety of other treatments. These procedures are prescribed seasonally as a preventive measure to ensure that all the doshas are kept in proper balance.

In 1983, the first Ayurveda clinic, known as Maharishi Ayurvedic Prevention Center, was established in Fairfield, Iowa, to offer panchakarma therapy as well as other treatments to the general public. Subsequently, other centers, known as Maharishi Ayurveda Medical Centers, have been established in Los Angeles, Cambridge, and Washington, D.C., as well as many other cities throughout the U.S. (see Appendix II). These centers have provided a setting to scientifically evaluate the benefits of Ayurvedic procedures, including panchakarma therapy. In one study, a self-report questionnaire was given to an experimental group undergoing panchakarma treatment and to a control group who knew of the benefits of panchakarma but did not undergo treatment. The experimental subjects reported significant improvements in a number of areas, including increased energy and vitality, improved appetite and digestion, and increased strength and stamina (Schneider et al., 1985b).

In a second study, a standardized psychometric test, the Profile of Mood States, was administered to similar experimental and control groups; again, the experimental group reported significant improvements such as decreased depression and fatigue and increased vigor and energy (Schneider et al., 1985b). In one further study, biological aging was assessed in subjects undergoing panchakarma treatment, using a method described in the next chapter. The preliminary results showed a significant reduction of biological age (by an average of 4.8 years) after two one-to-two week treatments over an eight-month period (Stryker et al., 1985).

Finally, Ayurveda not only includes approaches to perfect health on the level of consciousness, physiology and behavior, but also includes collective health measures on the level of the environment. The most significant of these collective health measures is the group dynamics of consciousness, briefly mentioned in the introduction and extensively discussed in chapter 8. Ayurveda is a very complete system of health which, if properly understood and researched, could be introduced inexpensively throughout the world.

On January 12, 1986, Maharishi inaugurated the World Plan for Perfect Health in order to create self-sufficiency in health care for every nation. The plan calls for introducing research and training programs so that the traditional systems of medicine in each country can be revived in the light of Ayurveda (see Appendix I). The foundation of the plan is the two most common and inexpensive resources of any nation: (a) the nervous systems of its people, brought into balance and unfolded for their full potential through the Maharishi Technology of the

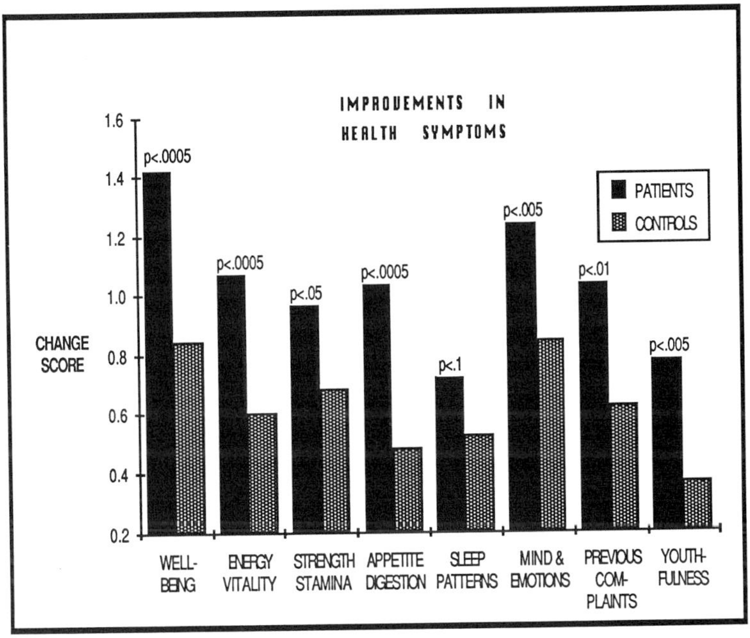

Figure 1. **Improvements in Health Symptoms - Health Benefits of the Maharishi Ayurveda Prevention Program.** This study compared l57 patients who participated in the Maharishi Panchakarma Program for one week of treatment to 65 controls who participated in a didactic class only and found statistically significant health improvements in the Ayurveda patients.

Reference: *R.H. Schneider, H.S. Kasture, S. Rothenberg, R. Averbach, K. Cavanaugh, R.K. Wallace: Improvements in Health with the Maharishi Ayurveda Prevention Program. Paper presented at the Eighth World Congress of the International College of Psychosomastic Medicine, Chicago, IL, 7 Sept. l985.*

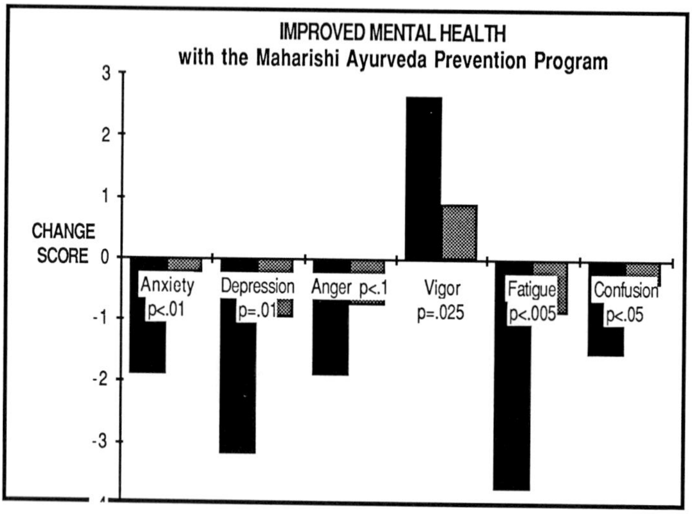

Figure 2. **Improved Mental Health with the Maharishi Ayurveda Panchakarma Program [MAPP].** This study tested 62 patients and 71 control subjects before and after the MAPP or a control condition. The results showed statistically significant declines in negative moods and improvement in the positive area [vigor].

Reference: *R.H. Schneider, H.S. Kasture, S. Rothenberg, R. Averbach, K. Cavanaugh, R.K. Wallace: Improvements in Health with the Maharishi Ayurveda Prevention Program. Paper presented at the Eighth World Congress of the International College of Psychosomastic Medicine, Chicago, IL, 7 Sept. 1985.*

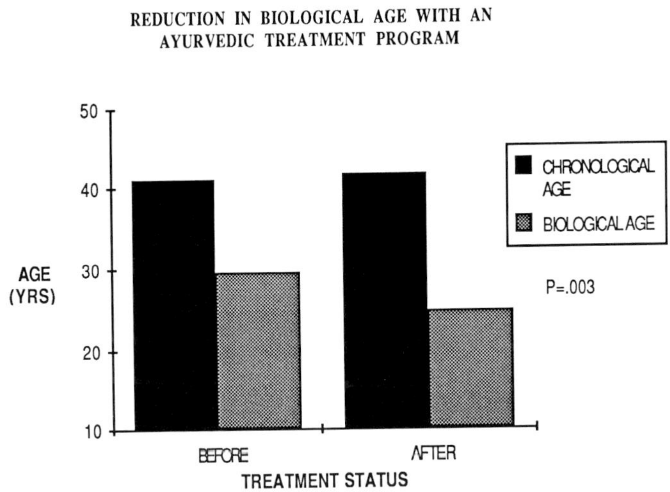

Figure 9. **Reduction in Biological Age with an Ayurvedic Treatment Program.** The Morgan Adult Growth Examination is a previously validated index of biologic age which includes systolic blood pressure, auditory threshold, and nearpoint vision. Ten subjects were tested with this battery before and after a 10-month period which included two Maharishi Ayurveda Panchakarma treatments of 1-2 weeks each. The results showed a decrease in biological age of 4.8 years.

Reference: *T. Stryker, R.K. Wallace, S. Rothenberg, R. Averbach. Paper presented at the Eighth World Congress of the International College of Psychosomastic Medicine, Chicago, IL, 7 Sept. 1985.*

Unified Field; and (b) the plants and minerals naturally growing in that geographical area, used for their medicinal values through the knowledge contained in Ayurveda.

Maharishi's World Plan for Perfect Health is ideally suited for developing nations that cannot afford the high cost of imported modern medical systems and are concerned about the many bad side effects of modern medicines. It is also suited for developed nations, since it provides profound knowledge of effective and inexpensive approaches for prevention.

The emphasis in modern medicine has been primarily on disease rather than health. More recently an increasing emphasis on preventive medicine has developed. The application of Ayurveda, expecially the Maharishi Technology of the Unified Field, for individual and collective health will bring fulfillment to health care systems around the world.

Maharishi, in his book *The Science of Being and Art of Living*, emphasizes the importance of immediately applying this effective formula for prevention of disease and elimination of suffering.

> A formula exists to take care of the very root of individual life, to maintain and restore health in all levels of mind, body and surroundings. We offer this formula in no spirit of competition or challenge, but out of love for man and with all good will towards those people all over the world who are endeavoring to alleviate suffering by whatever means they have found useful.
>
> To build more hospitals to alleviate sickness and suffering for people who have already fallen ill is a laudable act

of charity. But it is infinitely more important to find ways and means of preventing people from falling sick and of ensuring that they will always enjoy good health. Since a way does exist of preventing people from becoming sick, the introduction of it to people all over the world would be a much greater act of charity.

Chapter 7

REVERSING THE AGING PROCESS— LIFE IN THE DIRECTION OF IMMORTALITY

Why do different people age at different rates and in different ways? Today there are well over 10,000 men and women in the U.S. who are one hundred years or older. Why are these individuals different from the general population, whose average life expectancy extends to their early seventies? Are they more capable of resisting internal and external disorder and stress? Is their longevity due to their genes, their environment, or their way of life?

A number of now classic longitudinal studies have clearly described the effects of aging on physical and mental functions. These investigations, along with numerous other cross-sectional studies, report gradual but distinct changes in a variety of characteristics, such as standing height, weight, and skinfold thickness, and in specific physiological and psychological functions (e.g., basal and maximal oxygen consumption, blood pressure, vital capacity, reaction time, auditory thresholds, and specific measures of intelligence). These studies show that in general, different people age at different rates, and within one individual, different organs and systems show different rates of aging or loss of optimal operating ability.

The study of aging provides a very clear opportunity to test our hypothesis that more optimal states of physiological functioning exist and that perfect health and longevity can be cultured through the Maharishi Technology of the Unified Field as well as the various approaches of Ayurveda.

One of the first statements in the section of Vedic Literature on Ayurveda is:

> *Athato dirghajivitiyam*
> *adhyayam vyakhyasyamah*

> We shall now expound the chapter on the "Quest for Longevity."

The Vedic texts also state: "Ayurveda amritam"—Ayurveda is for the seekers of perfect health and immortality.

The entire purpose of Ayurveda is to cultivate that style of functioning of the body and nervous system which will develop and maintain higher states of consciousness. In the highest state of consciousness, enlightenment, one's awareness is established in pure consciousness, the field of immortality. Operating from this level, life is lived spontaneously in accord with natural law, and the value of that immortal level of life begins to be maintained in the physiology.

In this chapter we will concern ourselves specifically with the Ayurvedic approach to longevity on the level of consciousness and consider the effects of the Transcendental Meditation and TM-Sidhi program on various aspects of biological aging.

LONGEVITY FACTORS

Most researchers agree that, especially in technologically advanced countries, people rarely die of "old age." Long before they reach their full potential they succumb to the major killers of our era, such as cardiovascular disease and cancer. These diseases have been increasingly linked to the stress of modern life. To reverse the effects of aging, then, may not be as complex as it seems, for what we commonly understand to be "aging" may actually be the effects of stress. We have seen that the TM and TM-Sidhi program provides a very natural and easy means of reducing the harmful effects of stress on health and aging. This is perhaps most evident when we examine those factors that are most highly correlated with longevity.

Studies on aging by Palmore and his associates (1974) at Duke University have correlated the following seven factors with influencing longevity (listed in order of significance): (a) cardiovascular disease, (b) work satisfaction, (c) cigarette smoking, (d) physical function, (e) happiness rating, (f) self health rating, and (g) performance IQ. Studies on TM participants have reported improvements in all seven factors as we have noted in previous chapters. This research may be briefly summarized as follows:

(a) Studies show that a number of cardiovascular disease risk factors such as high blood pressure and high cholesterol levels improve as a result of the regular practice of the TM technique. (b) Job satisfaction also has been reported to improve with the TM technique. (c) Cigarette smoking and alcohol consumption decrease in TM participants. (d) We noted a variety of improvements in physical function, such as faster reaction time, improvements in respiratory and circulatory functions, and improved sensory performance. (e) Studies have shown increased happiness ratings, contentment, self actualization, and self-regard, and decreased anxiety and neuroticism. Increased happiness is further demonstrated by the significant improvement in hospitalized patients suffering from severe depression. These findings are also of interest in light of a study by Vallant (1979), which clearly establishes that better mental health is associated with longevity. (f) Improvements in self-health rating have been reported as well as (g) improvements in intelligence.

PROMOTION OF LONGEVITY

IMPROVEMENTS IN THE FACTORS MOST CLOSELY RELATED TO LONGEVITY

Longevity Factors		Improvements Occurring as Result of the Transcendental Meditation and TM-Sidhi Program
1	CARDIO-VASCULAR HEALTH	• Decreased hypertension • Improvements in patients with angina pectoris • Decreased cholesterol level • Increased cardiovascular efficiency • More efficient distribution of blood flow
2	WORK SATIS-FACTION	• Increased job satisfaction • Improved relationships with supervisors and peers • Improved job performance • Increased ego strength • Increased self-esteem • Increased self-actualization • Increased self-reliance • Improved work attitude
3	POSITIVE HEALTH HABITS	• Decreased cigarette smoking • Decreased alcohol consumption • Decreased use of non-prescribed drugs • Normalization of weight
4	PHYSICAL FUNCTION	• Improved auditory thresholds • Enhanced perceptual ability • Faster reaction time • Improvements in patients with bronchial asthma • Increased vital capacity • Improved running speed and agility • Improved perceptual-motor performance • Reduced inflammation • Relief from insomnia and improved quality of sleep • Faster recovery from stress • Improved biochemical stability, homeostasis, and efficiency • Reduced cortisol (a major stress hormone) • Restful style of physiological functioning
5	HAPPINESS RATING	• Increased happiness • Increased contentment • Increased self-regard • Increased adaptability to life events • Growth of inner well-being • Improvements in moods: increased relaxation, activation, and elation • Increased enthusiasm and dynamism • Growth of a more integrated, self-sufficient personality
6	SELF HEALTH RATING	• Improvements in self health rating • Reported ability to discontinue use of narcotics, tranquilizers, and sedatives • Reduction in reported use of non-prescribed drugs (in addition to improvements in objective measures of health)
7	INTELLI-GENCE	• Increased intelligence • Increased creativity • Increased learning ability • Improved academic performance • Increased orderliness of thinking • Improved concentration, comprehension, and memory • Increased innovation • Greater field independence (broader comprehension and improved ability to focus attention)
8	MENTAL HEALTH	• Improvements in psychological health • Decreased anxiety • Decreased neuroticism and depression • Decreased hidden mental turbulence • More stable internal frame of reference • Increased emotional stability and maturity • Improved social and psychological attitudes • Improved moral reasoning • Improvements in mood and body sensations • Growth of integration of personality

The eight factors shown above have been found to be most closely related to longevity. The research results corresponding to each factor, showing improvements in all these areas, indicate the stabilization of health and the promotion of longevity through the Transcendental Meditation and TM-Sidhi program.

The first seven factors shown above, in order of importance, were identified by a major study (E. Palmore, *Normal Aging II*, Durham: Duke University Press, U.S.A., 1974) to be closely related to longevity. The last factor, mental health, has been found to be of great importance for longevity and good health in later life (G. E. Vaillant, *New England Journal of Medicine*, 301: 1249–1254, 1979).

REVERSAL OF AGING

SCIENTIFIC RESEARCH INDICATING THE REVERSAL OF AGING THROUGH THE TRANSCENDENTAL MEDITATION AND TM-SIDHI PROGRAM

All the factors below show deterioration with the aging process as indicated by the arrows in the left column.

A reversal of the aging process is observed in every case as a result of the Transcendental Meditation and TM-Sidhi program as indicated by the arrows in the right column.

		Usual changes with age	Improvements with the TM and TM-Sidhi program
PHYSIOLOGY*	Blood pressure — systolic	↑	↓
	Blood pressure — diastolic	↑	↓
	Cardiovascular efficiency	↓	↑
	Cerebral blood flow	↓	↑
	Vital capacity	↓	↑
	Maximum voluntary ventilation	↓	↑
	Auditory threshold	↑	↓
	Auditory discrimination	↓	↑
	Near point of vision (accommodation)	↑	↓
	Homeostatic recovery	↓	↑
	EEG alpha power	↓	↑
BIOCHEMISTRY*	Serum cholesterol concentration	↑	↓
	Hemoglobin concentration	↓	↑
PSYCHOLOGY	Susceptibility to stress	↑	↓
	Behavioral rigidity	↑	↓
	Learning ability (paired associate learning)	↓	↑
	Memory — (short-term)	↓	↑
	Memory — (long-term)	↓	↑
	Organization of memory	↓	↑
	Creativity	↓	↑
	Abstract reasoning	↓	↑
	Intelligence	↓	↑
	Dichotic listening	↓	↑
MIND-BODY COORDINATION	Reaction time	↑	↓
	Sensory-motor performance (mirror star-tracing test)	↓	↑
HEALTH	Cardiovascular disease	↑	↓
	Hypertension	↑	↓
	Asthma (severity)	↑	↓
	Immune system efficiency	↓	↑
	Sleep disturbance (arousals/night)	↑	↓
	Insomnia and insufficient sleep	↑	↓
	Daytime sleep	↑	↓
	Quality of sleep	↓	↑
	Depression	↑	↓

All these factors show deterioration with the aging process. The opposite changes— indicating a reversal of the aging process—are observed with the Transcendental Meditation and TM-Sidhi program.

* A small number of changes resulting from the Transcendental Meditation and TM-Sidhi program are not directly opposite to the changes produced by aging. These changes are nevertheless important. For instance, cortisol levels and metabolic rate decrease with Transcendental Meditation and with aging. The changes occuring with aging are related to the accumulated loss of tissue as a result of general wear and tear, whereas the changes occurring with Transcendental Meditation are associated with decreased stress levels and greater efficiency of metabolic processes.

The above studies provide strong circumstantial evidence for the beneficial effects of the TM technique on the aging process. Let us now examine studies concerned with specific measures of biological aging.

BIOLOGICAL AGING

In order to more directly measure these effects, my coworkers at Maharishi International University and I have conducted experiments on the effects of the TM and TM-Sidhi program on biological aging (Wallace et al., 1982a). We tested the following two hypotheses: (1) that participants on the TM program would be biologically younger than norms for the general population, and (2) that this measure would be greater in individuals who have been practicing the TM technique for a longer period of time.

A number of researchers have attempted to develop instruments that would measure biological age effectively. These instruments have involved varying degrees of complexity, ease of administration, and utility of results. In the first of our studies we used the Adult Growth Examination developed by Morgan (Morgan and Fevens, 1972) as a measure for assessing biological age, because of its practicality and reported reliability. We administered the Morgan test to a cross-sectional group of short and long-term participants in the TM and TM-Sidhi program and to non-meditating controls.

The procedures and norms of the Adult Growth Examination were derived in part from the United States National Health Survey, which includes a carefully selected representative cross-sample of several thousand adults, and in part from a number of independent studies. The test has been validated in studies conducted in the United States and Canada and includes three basic subtests—measurements of auditory threshold, near-point vision, and systolic blood pressure—which Morgan has reported to be the most reliable and easily measured indicators of biological aging. These three subtests were therefore selected for this study.

Subjects were recruited from university staff and visitors and from the community through an article on biological aging in the local newspaper. Volunteers between the ages of 40 and 64 were studied. Eighty-four subjects, 38 men and 46 women, with a mean age of 53 years, participated in the study.

The controls included 11 subjects (5 men and 6 women) with a mean age of 54.2 years who did not practice the TM program. The remaining 73 subjects, 33 men and 40 women, acted as the meditating experimental group. Their mean length of TM practice was 62 months, ranging from 1 week to 181 months. The experimental group was divided at the approximate mean length of TM participation (5 years) into a short-term group (n = 33) and a long-term group (n = 40). The mean lengths of TM participation for these groups were 34 and 85 months, respectively. The short-term group consisted of 15 men and 18 women with a mean age of 52.2 years, while the long-term group consisted of 18 men and 22 women with a mean age of 53.3 years. The mean ages of the three groups were thus not significantly different.

The TM sample included persons practicing the TM-Sidhi program. Eleven of the short-term participants and 22 of the long-term participants had learned some or all of the TM-Sidhi program. Those in the short-term TM group had either recently completed, or had partially completed TM-Sidhi instruction. Practice of the TM-Sidhi program in the long-term group ranged widely from partial training to two years of experience.

The results showed that there was a significant difference between biological and chronological age in the long-term as compared to the short-term meditators and controls. The mean age difference score for the control subjects was -2.2 years, for the short-term TM sample -5.0 years, and for the long-term TM group -12.0 years.

The results indicated that the long-term TM meditators had significantly younger biological age than the short-term meditators, controls, and norms for the general population, and that the strength of this effect is related to the length of practice of the TM technique. Further statistical analysis indicated that while subjects who excluded red meat from their diets had younger biological ages, the effect of the TM program was independent of both diet and exercise pattern.

Two studies conducted in England have replicated and extended these initial findings. In one cross-sectional study TM subjects were found to have a biological age seven years younger than their chronological age. A follow-up study one-and-one-half years later found that the biological age of the same meditators actually decreased by another one-and-one-half years (Toomey et al., 1983a, Toomey et al., 1983b).

Another important recent study on biological aging involves the measurement of one of the most reliable biochemical markers of the aging process, serum dehydroepiandrosterone sulfate (DHEAS) by Glaser and coworkers (1986). DHEAS declines progressively with age. Peak levels occur in one's mid-twenties; by the eighth and ninth decades of life, one's DHEAS level may have declined 80 percent.

DHEAS levels were measured in 254 men and 74 women who were experienced participants in the TM and TM-Sidhi program, and compared according to sex and five-year age groups to 981 men and 481 women control subjects who were not meditating. The mean DHEAS levels in meditators were significantly higher than in the controls in all five of the age groups measured in women,

Reduction of Biological Age

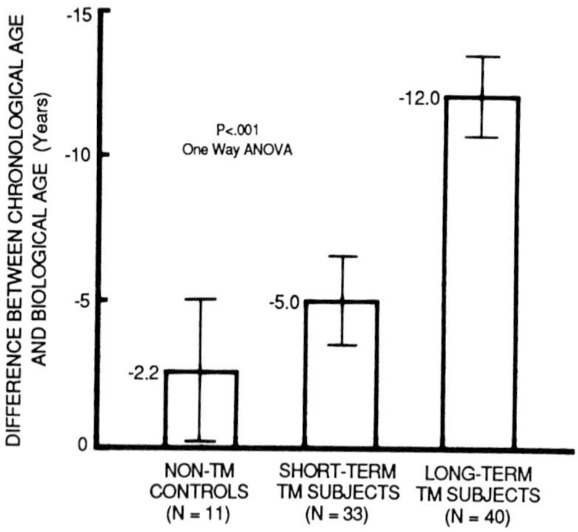

A standardized test of biological age, the Adult Growth Examination, was given to a cross-sectional group of men and women with a mean age of 53 years. They were divided into three groups: nonmeditating controls, short-term TM meditators (under five years) and long-term TM meditators (over five years). The average biological age of the nonmeditating controls was 2.6 years younger than their chronological age; of the short-term meditators, 5.0 years younger than their chronological age; of the long-term meditators, 12.0 years younger. A statistically significant difference was found between the long-term meditators and both the control group and short-term meditators ($F = 8.05$, $p < .001$, one-way ANOVA). Further, a significant correlation ($r = .46$, $p < .001$) was found between younger biological age and length of time practicing the TM technique.

Reference: R. Keith Wallace, Eliha Jacobe, and Beth Harrington, " The Effects of the Transcendental Meditation and TM-Sidhi Program on the Aging Process." *International Journal of Neuroscience*, 16 (1) 1982: 53-58.

Reduction of Age in a Longitudinal Study

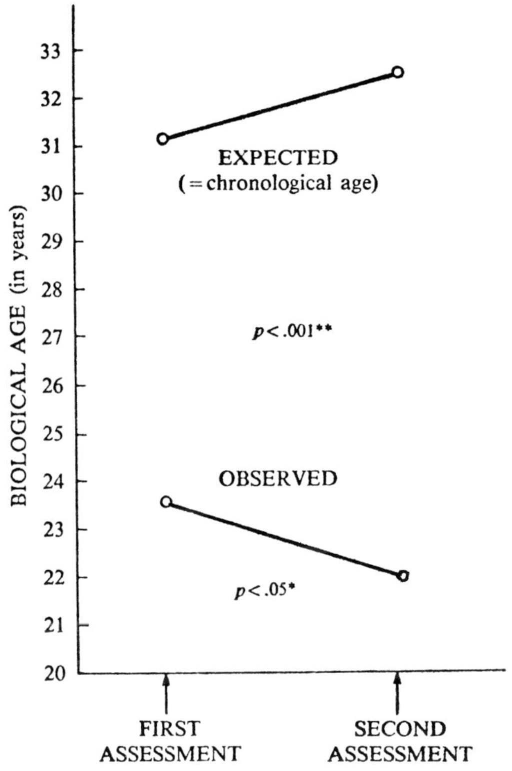

* paired t-test, for change in observed biological age.
** paired t test, for change in difference between observed and expected biological age.

Mean biological age decreased by 1.7 years during the interval of study ($p<.05$). There was therefore a significant increase in the degree to which biological age was younger than chronological age, the difference being 7.5 years at the first assessment and 10.6 years at the second assessment ($p<.001$).

TOOMEY, M.; PENNINGTON, B.; CHALMERS, R.; and CLEMENTS, G. The practice of the Transcendental Meditation and TM-Sidhi programme reverses the physiological ageing process. MERU Research Institute, Mentmore, Buckinghamshire, England, and Department of Biology, University of York, Yorkshire, England, 1982.

and in eight of eleven age groups in men. Further, the difference was more pronounced in the older subjects: the mean percentage of DHEAS elevation over control values was higher for the older age groups, with mean differences of 23 percent for men over 45, and 47 percent for women over 45. This effect was independent of identifiable contributing factors including diet, exercise, obesity, or use of alcohol.

The mean levels measured in the older meditators were generally comparable to levels in control groups five to ten years younger. Low levels of DHEAS in women have been correlated with higher incidence of breast cancer, and administration of DHEAS or DHEA has been shown to improve obesity, diabetes, spontaneous and induced tumors, and autoimmune processes in mice—all diseases associated with senescence. That regular practitioners of the Transcendental Meditation technique have higher DHEAS levels than control populations is another indication of younger biological age in TM participants.

One further recent study considered the effects of TM on cognitive and behavioural flexibility, health and longevity in elderly individuals (Alexander et al., 1983). In assessing the effects of the TM program on aging the study examined whether such effects are produced independently of context, expectation, or simple relaxation components.

Seventy-three residents of homes for the elderly (60 women and 13 men, with a mean age of 80.7 years) were randomly assigned to either a no-treatment condition or to one of three treatments designed to be equivalent in external structure and features fostering expectation: the TM program, an active thinking procedure (mindfulness training), and a relaxation program. All groups were initially similar in expectancy of benefits and on pretest measures, yet after a three-month experimental period the TM group had significantly improved in comparison to one or more treatment conditions on three measures of cognitive flexibility (the Overlearned Verbal Task, the Stroop Color-Word Interference Test, and the Associate Learning Test for difficult word pairs), word fluency, two methods of assessing change in systolic blood pressure, self-report measures of behavioural flexibility and aging, and in nurses' rating of mental health (after 18 months). Also, the TM subjects reported feeling more interested during their practice, and better and more relaxed immediately after their practice, than did the active thinking and relaxation subjects. Overall, more TM subjects found their practice to be personally valuable than did members of the other treatment groups. The active thinking group scored higher on a self-report measure of internal locus of control and, like the TM group, improved in word fluency, mental health and, to a lesser degree than TM, on a measure of systolic blood pressure.

The most striking finding is that in addition to reversing age-related declines, TM appears to directly enhance longevity: all members of the TM group were still alive three years after the program began, in contrast to the other groups and to the 62.5 percent survival rate for the remaining population in these homes for the elderly (n= 478).

The results of these studies on biological aging, supported by the large number of other physiological studies, strongly indicate that the Maharishi Technology of the Unified Field retards and even reverses the aging process.

Two distinct types of underlying physiological mechanisms that might account for these long-term changes are suggested by the previous research: (1) an increased ability to maintain a lower, more stable, internal resting state, and (2) an improved, more adaptable response to external stress. Let us examine each of these two mechanisms in detail.

MORE STABLE AND EFFICIENT INTERNAL PHYSIOLOGY

The research findings, as described earlier, indicate that during the TM technique participants attain a state of deep rest, which in terms of physiological functioning is opposite to that of the stress response. In addition, there is evidence from the studies we have already discussed that certain physiological and biochemical values are lower in TM meditators, even when they are not practicing the TM technique, than in non-TM meditating controls. These values include: spontaneous skin resistance fluctuations, blood pressure, respiration rate, heart rate, pituitary hormone levels, and oxygen consumption. The increased ability of TM participants to maintain a more stable and restful physiological state simultaneously increases their performance on reaction time tests, perceptual-motor tasks, and job productivity. Thus, these findings may be interpreted to mean that the physiology of TM meditators operates in a smoother, more orderly and more efficient mode.

The more restful functioning of specific physiological parameters in TM participants is particularly interesting in the light of animal research. Bourliere (1970) reviews the long-standing observation that low body metabolism is definitely correlated with longer life span in animals. As Bourliere notes, Rubner (in 1890) was the first to propose a definite relationship between the rate of metabolism and duration of life. It was his theory that various mammals during their lifetimes used up approximately the same number of calories per unit weight. Smaller animals, Rubner reasoned, in order to preserve the constancy of their body temperatures, had to have a much higher rate of metabolism per unit weight than large ones (for example, the metabolic rate of a harvest mouse is 2.50 liters O_2 /mg/hr, while for an elephant it is 0.07 liters O_2 /mg/hr). They also live a much shorter life, three to four years for a mouse as compared to the elephant's 70 years. Bourliere also points out that other physiological parameters such as temperature and growth rate, which are known to affect metabolic rate, can also influence the life span of an animal.

The research by McCay at Cornell University demonstrates one of the most exciting and yet least understood effects on longevity, which may relate to the importance of maintaining a stable and restful physiological state. McCay's original experiments were published over 50 years ago and have been since replicated by numerous researchers (McCay and Crowell, 1934; McCay, 1952; Ross and Bras, 1975).

In one experiment he randomly assigned a litter of rats either to an experimental group that received only essential proteins, minerals and vitamins, or a control group that were allowed to feed as they chose with an unlimited supply of food, including carbohydrates and fats as well as the essentials. The control group averaged a normal life span. The experimental group, however, lived significantly longer, with the oldest ones living in general twice as long as the controls—the oldest rat lived 1,320 days (about 132 human years equivalent). McCay also observed that the experimental rats at a very old age maintained a healthy appearance and engaged in vigorous activity, and that when they were allowed to eat at will, they experienced a spurt in growth.

This research suggests that animals that conserve more energy through lower, stabler, and less entropy-producing physiological functions live longer and have better health at older ages. Thus, the ability of TM meditators to achieve a lower and more stable and orderly resting state both during the TM technique and during activity might be a very important mechanism in achieving and maintaining a younger biological age.

The second important mechanism mentioned above

relates to how TM participants adapt and recover from challenges to their systems. The ability of a living system to adapt and maintain internal stability in the midst of external change generally decreases with aging. Research on TM subjects, as we have outlined, suggests a definite improvement in homeostatic mechanisms. These and other beneficial effects of the Maharishi Technology of the Unified Field described in chapter 6 should have an important influence on retarding or reversing the aging process.

Let us briefly examine some of the major theories of aging and attempt to develop a detailed and complete model for the effects of the TM and TM-Sidhi program on the aging process.

THEORIES OF AGING

The study of aging is one of the newest and most expanding fields in modern science. Perhaps as a result of its interdisciplinary nature, it is comprised of an increasing number of divergent and specialized theories. Attempts to categorize these many theories have crystallized into two basic themes which suggest that aging is due either to (1) a predetermined genetic program or "clock", or (2) the accumulation of errors resulting in functional or structural abnormalities at various levels of physiological organization, caused either by extrinsic or intrinsic factors.

In regard to the first theme, there are many researchers who feel that the damage or disorder seen in older organisms is not a result of a random or accidental process, but is rather due to a specific genetic program. For example, Bernard Strehler, one of the outstanding gerontological researchers, believes that the process of aging is similar to other stages of development and is controlled by a specific set of "aging genes." According to this theory, when the "aging genes" in the DNA molecule are activated they then regulate the gradual breakdown of the organism. Strehler (1977) points out one of the important implications of this theory: that it may be possible through a very refined genetic technology to either halt or reverse the aging process.

In regard to the second theme, many researchers feel that aging occurs as the result of the inability of the organism to properly repair errors that accumulate over time. The precise cause of the accumulation of errors and the level of physiological organization at which their effect is most important differ according to different researchers.

At the molecular and cellular level, the somatic mutation, redundancy, error "catastrophe," cross-linkage, and free radical theories have been proposed (Finch and Hayflick, 1977). It is not the purpose of this book to explain these theories in detail. Let us consider briefly, however, an interesting and comprehensive theory by the late George Sacher.

Sacher (1978) postulated the existence of "longevity genes" (or "longevity assurance genes") as opposed to "aging genes," and expressed the various error theories in terms of an evolutionary perspective. Sacher's proposal is unique within error theories in that it emphasizes the genetic ability to repair errors rather than the causes of the errors themselves. Sacher reasoned that the direction of evolution is towards the acquisition of specific genes (the "longevity genes") that enable the development of more and more complex self-repair and self-regulatory systems.

Research by Sacher and others showed that the more highly evolved mammals have both more sophisticated DNA self-repair mechanisms and more complex nervous systems, both of which are essential for eliminating errors or instability within an organism. Sacher had found that the physiological parameters most highly correlated with longevity in different species are lower metabolic rate and larger brain size compared to body size. He concluded that these properties are related to more efficient metabolism and more complex nervous systems in higher mammals, and to the evolution of longevity genes which, by virtue of providing greater flexibility in dealing with its environment, enable the animal to live longer.

Sacher's research and theories provide a very rich model that helps to explain how the Maharishi Technology of the Unified Field could extend life in the direction of immortality. They further suggest a number of interesting experiments to test the effect of the TM and TM-Sidhi program on the aging process—for example, testing the DNA repair efficiency of cells taken from long-term TM meditators as compared to control subjects.

Maharishi has suggested for a number of years that the effects of the TM and TM-Sidhi program on DNA and the mechanics of its expressions should be studied. According to Vedic knowledge, the refinement of consciousness develops the full potential of all levels of physiological functioning. The Technology of the Unified Field, as described by Maharishi, establishes our awareness in the least excited state of consciousness, that most basic level of energy and matter, which is the ground state of natural law. From this level it should be entirely possible to directly influence genetic information and machinery and thus unfold the full potential of the DNA for living life in the direction of immortality.

What mechanisms might exist for unfolding more of the potential of DNA? In considering this question let us examine several further theories of aging. The mechanisms by which DNA regulates aging have been suggested by several researchers, with reference to gene expression in specific systems of the body such as the immune system or the neuroendocrine system. For example, Paola Timiras at the University of California at Berkeley emphasizes the role of the neuroendocrine system in aging. Timiras (1978) suggests that, whereas differences in lifespan and rate of aging depend upon intrinsic (genetic) and extrinsic (environmental) factors, aging itself is an inevitable developmental process controlled by some type of pacemaker center in the brain.

Ultimately, this pacemaker would be structured by the information in DNA, but its expression would be localized to certain cells within the brain. Timiras suggests that this pacemaker center, acting via neuroendocrine mechanisms—specifically the pituitary gland—causes progressive change throughout the body in a manner similar to that seen in other stages of development, such as birth or adolescence. She also suggests that despite the complexities involved, the aging brain may be explained overall in terms of chemical imbalances in the brain.

The imbalance of chemicals, specifically neurotransmitters, would have repercussions not only on endocrine function but also on autonomic, sensory, motor, and behavioural functions. As Timiras points out, we already know that normal brain functioning depends upon a critical balance of neurotransmitters; any imbalance has widespread implications. In the aging brain the situation may be even more critical, and even minor shifts in neurotransmitter balances might lead to significant functional repercussions. Preliminary research at her laboratory and others has shown that dietary and pharmacologic manipulation of brain neurotransmitters such as serotonin and dopamine in rats can influence their growth and development, delay age-related decrements in their homeostatic and reproductive function, and significantly extend their life span.

Caleb Finch of the University of Southern California, one of the foremost researchers in the field of gerontology, has also conducted a number of studies suggesting that aging may be influenced by neurotransmitter balance.

Finch (1977) has found that the activity of a group of neurotransmitters known as the catecholamines is reduced with aging, particularly in the hypothalamus. He feels that this reduction could markedly influence the secretion of hypothalamic releasing factors and thus affect many aspects of neuroendocrine function, for example the change in ovarian cycles, which occurs during midlife in many mammals.

The suggestion that neurotransmitter balance and changes in neuroendocrine function may be central to the process of aging is particularly interesting in the light of short and long-term studies on the effects of the TM and TM-Sidhi program on hormone and neurotransmitter levels, which we have discussed in the previous chapter. One particular finding, long-term decrease in TSH levels in the pituitary gland, is of especial interest since previous studies have reported a higher prevalence of elevated TSH levels in the elderly.

The Maharishi Technology of the Unified Field, by influencing neuroendocrine functioning and chemical balance in the brain, could cause many physiological changes. These changes may be caused by an alteration in the expression of the information in the DNA of certain pacemaker cells in the brain. We will explore this possibility more in the following section, in the light of current research on the role of experience on the development and aging of the brain.

THE POTENTIAL TO OPPOSE AGING IN THE BRAIN

Marian Diamond, of the University of California at Berkeley, in an article entitled, "The Aging Brain: Some Enlightening and Optimistic Results," suggests that in "the absence of disease, impoverished environment, or poor nutrition, the nervous system apparently does have the potential to oppose marked deterioration with aging."

(Diamond, 1980) Her own studies, along with a number of other experiments in the field, show that experience can markedly affect both development and aging in the nervous system (Bennett er al., 1964; Diamond et al., 1976; Buell and Coleman, 1979).

In a series of experiments, rats were raised in two different environments, one that was "rich"—containing many different types of stimuli and opportunities for perceptual motor activity—and one that was "poor"-a standard small wire cage. The effects of the different environments on the brain were very noticeable. The rats living in the rich environment developed bigger and heavier brains, with a thicker cerebral cortex. There were also chemical differences that suggested a greater quantity of a key neurotransmitter, acetylcholine. Perhaps most interestingly, the nerve cells in the cerebral cortex of the rats in the enriched environment had more dendritic spines (small thorn-like protrusions associated with synapses), and thus presumably a greater number of structural neural connections, than the rats in the poor environment.

Diamond and coworkers conducted aging experiments in which the external environments of the rats were not impoverished but kept at a standard level through their lifetime (Conor et al., 1980). A number of neural parameters (such as neuron number, density, and size, as well as the number of glial cells in the occipital cortex) showed their greatest decline in the initial 108 days of life, but from then until 542 days, when the rats were very old, there was little or no change at all in the neural parameters. The importance of Diamond's and coworkers' findings is that understanding those factors that affect neural development may give us the ability to both develop more of the inherent potential of the brain and retard the detrimental effects of aging. With this in mind, let us now look more closely at the detailed mechanics of the process of development in the nervous system.

DEVELOPMENT OF THE NERVOUS SYSTEM: THE IMPORTANCE OF EXPERIENCE

What controls the growth of a neural process? How does it know where to go and with which other nerve process to form a synapse? Is it genetically programmed or is experience an important factor? How fixed are the connections? Can they be changed? These are among the many important questions that are being researched in developmental neurobiology today.

Studies on less evolved animals have shown remarkably constant neuronal geometry; an expert in the field can readily identify the same specific neuron from one animal to another among the same species. The use of special dyes has enabled researchers to map the whole structure of a given neuron and its processes, again revealing a remarkable constancy of major connecting pathways among the same species.

The research thus documents that precise genetic programming is a very prevalent and important aspect of nerve growth. It also highlights the relative importance of genetic programming in most animals. For example, only moments after its birth, the African wildebeast is able to get up and run with the herd. Obviously, that animal's nervous system must be nearly fully developed at birth. On the other hand, the human infant is almost helpless at birth; the nervous system must undergo years of experience before perceptual-motor coordination is developed.

A famous set of experiments demonstrating the importance of experience was conducted by Hubel and Wiesel (1979) at Harvard University. Within the visual cortex of cats are extremely specific cells that tend to respond best to one type of stimuli—for example, some cells respond best to a horizontal bar located at a definite point in the visual field, while other nearby cells respond best to a vertical line in a particular location.

In order to determine the role of experience in establishing and maintaining the connections to these cells, Hubel and Wiesel conducted experiments in which they deprived newborn kittens of certain stimuli. If this deprivation took place during a critical period (which seems to last only three or four days during the fourth and fifth week after birth), dramatic changes occurred within the fine structure of the brain. There was a marked decrease in the number of dendritic connections to those cells genetically specialized toward the stimuli of which the animal was deprived.

In another experiment conducted by Blackmore and Cooper (1970) in England, kittens were raised in a highly modified environment, consisting of either vertical or horizontal bars, during their critical visual experience periods. When the cells in the visual cortex of each of these animals were tested, it was shown that they were profoundly affected by the lack of experience. The cells in kittens raised in a horizontally striped environment responded almost exclusively to horizontal stimuli and not to vertical stimuli. The connections to the cells responsive to horizontal stimuli were numerous, but because of the imbalanced experience, connections to vertical responsive cells had literally withered away.

REORGANIZATION OF NEURAL PATHWAYS IN HIGHER STATES OF CONSCIOUSNESS

These and other similar experiments concerning the importance of experience in the development of neural pathways have important implications in terms of the development of higher states of consciousness. What if an individual has never had the experience of allowing the mind to settle to quieter, more refined levels of consciousness, of transcending more excited mental states to the least excited state of consciousness—pure consciousness?

Will the structural connections of his nervous system be different from those in another individual who has had that experience? I think the answer is yes—there will be a marked difference in neural organization, which could have enormous implications in terms of the realization of the individual's full genetic potential, affecting not only the expression of his full mental and physical abilities but also his health and longevity. I suggest that the daily practice of the Transcendental Meditation and TM-Sidhi program, by periodically providing the unique subjective experience of unbounded bliss consciousness, creates a new chemical and electrical environment in the brain which in turn expresses new information from the DNA. This new information then results in the reorganization of the nervous system and may enable critical new neural pathways to be formed.

These pathways are presumably part of our genetic endowment. However, if we do not provide ourselves with the unique set of inner experiences that foster and strengthen those pathways, they are lost to us. We live in a society that is preoccupied with the horizontal, surface level of thinking, and with obtaining more and more information at the fastest possible rate. It is as if we have grown up in an environment with only horizontal lines.

The result is that we have few opportunities to enliven the vertical direction of thinking—to actually transcend and experience the basis of thinking, pure consciousness. The neural pathways that support the experience of higher states of consciousness have not been cultured, have not been reinforced by experience, and consequently have literally withered away. We have come to accept a very limited range of human potentialities, precisely because we have not had within our educational systems the technology for experiencing more powerful and comprehensive levels of thinking and thereby enlivening the neurophysiological pathways that support transcending and the unfoldment of full creative intelligence.

In a conference entitled "Science, Consciousness and Aging," in Seelisberg, Switzerland, in 1980, Maharishi emphasized the importance of the experience of unbounded bliss consciousness. The following excerpts from Maharishi's comments concern the role of experience in maintaining the liveliness of neural connections:

> The point of consideration is what experience is it which will will keep the connections lively and not exhaust them. Unbounded consciousness is a level of consciousness which doesn't change, which doesn't decay. Experience of unbounded consciousness becomes the mechanics of maintaining the entire physiology completely lively. One more element, which we find in that experience of unbounded consciousness, is the element of bliss. Since unbounded consciousness is bliss, the experience of it can never be exhausting. The experiencing process has to be such that the experience of unboundedness is maintained when the boundaries are being experienced. This skill in the art of experiencing will be the mechanics of disallowing the physiology to age.
>
> When we want every experience within boundaries to be accompanied by the experience of the unbounded then we want both of these in the most natural way—neither there should be strain on the physiology to maintain the experience of unboundedness nor should there be strain in the physiology for the experience within boundaries. There can be strain to maintain unboundedness if one is on the level of mood making. That would strain the

physiology because then we are making the physiology function on a purely imaginary level of unboundedness yet still within boundaries. It could be a very great strain and such a thing will not eliminate the aging process. The skill of maintaining the experience of unbounded bliss consciousness at the time when one is exposed to experience within boundaries should be a spontaneous phenomenon, it should be most natural, with no labor involved.

The Upanishads demand that this process of culturing take time into account. There is a verse in the Bhagavad-Gita which says, there is nothing more purifying than pure knowledge. And how is this pure knowledge gained? By exposing the awareness to transcendence and experiencing the infinite dynamism in the silence of transcendental consciousness. There it gets stabilized. It takes time for the stabilization of bliss consciousness, so that the bliss consciousness is not just a vacuum state of consciousness, not just a least excited state of consciousness, but is lively in its value of bliss. So for transcendental consciousness to keep its bliss character lively it needs time. Time is needed because the experiencing faculty is physical. It is the physiology that has to tune itself to that kind of functioning which will maintain bliss consciousness. It doesn't have to be an entire lifetime. It could be within two, three, four, five years, and this is quite a long time. The time is necessary. There is another verse in the Bhagavad-Gita that says that once it starts, there is no obstacle or hindrance to this process of gaining bliss

> consciousness.
>
> Once started in that direction, the awareness becomes more and more blissful, until it becomes bliss completely. In this process of culturing the physiology, the repeated experiences of bliss consciousness and the stabilizing of that experience occur in such a way that even when one is out of transcendence, bliss consciousness is carried in the same awareness that experiences the relative. It becomes a continuum. And this becomes the formula for disallowing aging, a formula to maintain the youthfulness of the nervous system on the physiological level.

What evidence do we have that a structural reorganization takes place in the nervous system as a result of the sustained experience of pure unbounded consciousness? In general, we could attribute all the innumerable changes that occur as a result of the Maharishi Technology of the Unified Field to the regular experience of pure consciousness. More specifically, several studies we have described have looked closely at the correlation between neurophysiological measurements and clarity of experience of pure consciousness.

For example, as we mentioned in chapter 5, Haynes and coworkers reported that subjects with clear experience of pure consciousness showed faster recovery of the paired Hoffman (H) reflex, and that this correlated significantly with higher EEG coherence and creativity measures. In both this study and the study by Orme-Johnson, the clarity of certain key states during TM, including transcendental consciousness, appeared to be a critical variable in predicting the magnitude of change in certain physiological and psychological variables.

EEG COHERENCE, CREATIVITY, NEUROLOGICAL EFFICIENCY AND THE EXPERIENCE OF TRANSCENDENTAL CONSCIOUSNESS

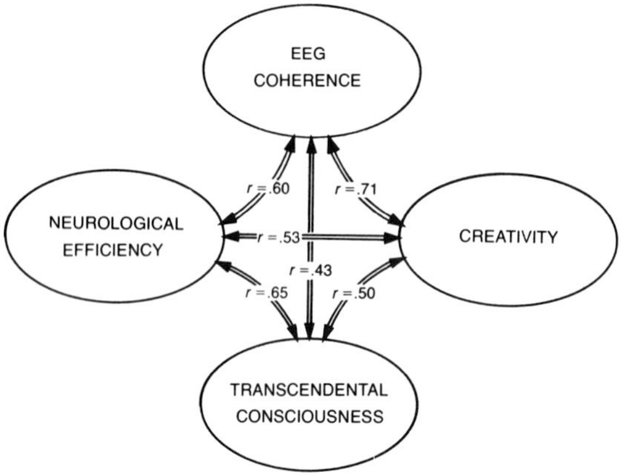

In a study of 23 individuals taking part in an advanced course in the Transcendental Meditation programme, positive and significant correlations were found among the following: EEG coherence during the TM technique, verbal creativity, speed of recovery of spinal motor neurons after stimulation (Hoffmann Reflex) and the clarity of experience of transcendental consciousness. This finding confirms that the growth of consciousness, which is reflected by the clarity of experience of transcendental consciousness, is accompanied by increased orderliness of brain functioning, and by increased efficiency of physiological and psychological processes

Reference: David W. Orme-Johnson, Geoffrey Clements, Christopher T. Haynes, and Keredine Badaoui, "Higher States of Consciousness: EEG Coherence, Creativity, and Experiences of the Sidhis" (MERU, 1977). Published in *Scientific Research on the Transcendental Meditation Program: Collected Papers*, Volume I, 2nd edition. West Germany, MERU Press, 1977.

As we also mentioned, in a longitudinal study on the H-reflex, the group that began the TM-Sidhi program showed a faster recovery of H-reflex responses and an increase in EEG coherence after only a three-to-six-month period. In addition, the faster recovery of H-reflex response was correlated significantly with clearer experiences of pure consciousness and with higher academic performance. Since H-reflex recovery in TM-Sidhi participants has been correlated with higher levels of creativity and better academic performance, and since numerous studies have demonstrated improved physical and mental functioning as a result of the TM technique, it seems clear that the regular experience of pure consciousness results in a more optimal style of neurophysiological functioning, one in which the nervous system is less biased by prior conditioning and more sensitive and accurate in its responsiveness.

Thus there appears to be good experimental evidence for the hypothesis that the experience of higher states of consciousness results in the reorganization or strengthening of certain connections and pathways which enable the brain to develop a more optimal and orderly style of functioning.

In comparative anatomy studies, as we have noted, the one factor that correlates most highly with longevity is the size of the brain in relation to the body. Sacher and others have suggested that it is the higher intelligence provided by a large, more complex brain that enables more highly evolved animals to live longer. Man, the longest-lived mammal, has enormous intellectual and creative capacities. If he were to expand those capacities to their full value, the highest level of consciousness, enlightenment, what then would his life span be?

According to Maharishi, in the state of enlightenment, in which perfect mind-body coordination is attained and the individual consciousness is established in the state of pure unboundedness—bliss consciousness—anything is possible. Maharishi explains that while it may be impossible to know all the laws of nature and their implications in every situation, it is entirely possible to be all the laws of nature. Operating from the level of pure consciousness, the individual is in perfect harmony with the laws of nature and spontaneously is sensitive to both his own needs and those of his environment.

This is extremely important, particularly in our age of technology. There is an old proverb, "A little knowledge is a dangerous thing." There is little question of the enormous repercussions that could result from technological manipulation without complete knowledge, without being in complete harmony with all the laws of nature. Between genetic engineering and nuclear power we live in a very precarious world. As we have gained deeper insight and knowledge into the laws of nature and developed the technology to tamper with these laws, the responsibility for proper and holistic judgement has increased beyond any expectation. Knowledge must give an evolutionary direction to power.

It is only by raising the level of collective consciousness, increasing the creative intelligence of all the individuals who share and control the world, and expanding science to include a perfect technology of consciousness, that we can begin to feel safe and confident in this nuclear age. To live life in the direction of immortality we must live life in accord with the laws of nature.

LIFE IN ACCORD WITH NATURAL LAW

There are many obvious and extremely important benefits in a society in which individual members behave in accord with the laws of nature and are thus able to live a long life in good health. Traditionally, the elders of any society have served as the custodians of wisdom and cultural integrity. It is through their living example and the depth and breadth of their knowledge that the stability and strength of the culture is maintained generation after generation. Their very presence inspires the youth of that society with a purposeful direction and goal. Their years of experience provide an enormous resource to the young, helping to guide and educate them to avoid unnecessary pitfalls and giving them the experience and expertise so necessary for wise and progressive leadership.

Perhaps one of the most deplorable situations in many modern western cultures today is the lack of respect and purpose the elderly are given. The whole structure of society has been turned upside down. Instead of naturally commanding the respect of the youth, the elderly are treated as outcasts, as weakened burdens to society. To remain a part of society and lessen the generation gap, they are often forced to emulate the appearance and action of the youth, and the value of their wisdom and experience is forgotten or not appreciated. The heritage of our cultural values is lost and traditional family structures are weakened from within.

The real seeds of this problem were sewn long ago with the loss of the technology to stave off deterioration and insure continued mental and physical development. Science and technology have disrupted and devastated most cultural traditions. Change and material progress have replaced the older values of family and community stability and cultural wisdom. The elders of any society, in order to retain their role as leaders and guides, have attempted to keep up with the rapid pace of change. Many have great difficulty integrating the values of their very rich, yet apparently outdated, traditions of the past with their present modern style of living. The result is often a crisis in physical and mental health in all levels of society. It is as if the society as a whole, because of its poor state of health and vitality, begins to age.

In the process of maturation within such a society, the

adults have had to neglect their traditional family roles, resulting in the loss of valuable guidance for the children and an over-all weakening of the family structure itself. Further, in order to cope with the stress of the ever-increasing demand to be successful (and the demands of success once it is achieved), they often succumb to excessive smoking and drinking, even though these are known to increase the likelihood of accident and disease.

What is necessary to reverse these negative trends of life? Each individual must have the inner technology to prevent illness and deterioration and to become increasingly more powerful and flexible every year of his life. Maharishi's revival of such a technology is of enormous value to the progress of mankind.

Research on the Maharishi Technology of the Unified Field and on Ayurveda clearly shows that the prospect of growing old need no longer signify a loss of life, but rather a time of greater refinement of abilities and real increase in power and happiness. Each year then becomes an opportunity for further development, for tapping new and richer creative potentialities, growing eventually to such a state of perfection and enlightenment that all possibilities are available. This is a state of life to be emulated by the young, a state of life worthy of the leaders of a society.

Individual consciousness is the unit of collective consciousness. When the individual, acting from the level of infinite correlation, becomes the embodiment of perfect health and longevity he establishes the full potential of natural law in society, and cultures perfect health and longevity at every level of collective consciousness: family consciousness, community consciousness, city consciousness, national consciousness and world consciousness. This is the meaning of perfect health and life in the direction of immortality: life in accordance with natural law on all levels—individual, social and cosmic.

Chapter 8

IS CONSCIOUSNESS A FIELD?

Generation after generation, men have planned and dreamed of establishing a utopia, an ideal world of peace and harmony where all people are united by a common bond of friendship and compassion and a common interest in progress and happiness. The history of individual groups, governments, and political systems in all countries clearly shows that no one has succeeded in establishing and preserving an ideal society. No matter how valid and appealing one system may appear, no matter how great the material and intellectual resources available, governments everywhere are burdened by problems and unattainable goals.

Is civilization thus doomed to conflict and aggression by the very nature of human behavior, as many famous psychoanalysts and sociologists have expressed, or are there deeper principles that have been overlooked, forgotten throughout the ages? According to Maharishi, only lack of knowledge concerning the real nature of human consciousness has prevented the emergence of an ideal society. In the following passage , Maharishi explains the situation clearly:

> The failure of defense, government, education, economics, ecology, law and order, and every other area of responsibility in society was not the fault of either the leaders or the people. The actions of the leaders of society and governments express the value of collective consciousness. All the leaders of society sincerely did what the

> collective consciousness drove them to do—they did what they were capable of according to the situation and circumstances. The people are also not at fault because they acted on the basis of whatever they were taught and the way of life shown to them.
>
> The failures are neither the fault of the people nor the fault of the leaders, the only fault was that ignorance prevailed and the infinite potential of man was not known. The situation is that no one was at fault and no one was wrong—only the right was missing.
>
> Men everywhere have always tried their best to bring security, success, and happiness to mankind, but the knowledge on the basis of which they were acting was inaccurate and incomplete. That knowledge belonged to the age of ignorance, which has now been replaced by the Age of Enlightenment. With the restoration of the knowledge of supremacy of consciousness as the prime mover of life, the age of ignorance is receding and completely new principles of life and living are emerging. The right is now being re-established because of the rise of world consciousness brought about by the people already practicing Transcendental Meditation.

Maharishi further explains that there exists a collective consciousness, which has its ultimately basis in the field of pure consciousness. The consciousness of each individual in a society contributes to the overall coherence of collective consciousness, which in turn influences the individuals and determines the character of a society.

When individuals have the knowledge and technology to directly experience the underlying field of pure consciousness, not only can individual behavior be changed, but more importantly the collective consciousness can be influenced and the behavior of the whole society can be improved. To understand the mechanics of this process let us first briefly examine the concept of collective consciousness.

COLLECTIVE CONSCIOUSNESS

The concept of a collective consciousness underlying and influencing the structure of society has been expressed by many great thinkers in the past. William James vividly describes this level of collective consciousness shared by all men and women in the following passage:

> Out of my experience, such as it is (and it is limited enough) one fixed conclusion dogmatically emerges, and that is this, that we with our lives are like islands in the sea or like trees in the forest. The maple and the pine may whisper to each other with their leaves and Connecticut and Newport hear each other's foghorns. But the trees also commingle their roots in the darkness underground and the islands also hang together through the ocean's bottom.
>
> Just so there is a continuum of cosmic consciousness against which our several minds plunge, as into a mother sea or reservoir.

Collective consciousness, however, has never been studied in a serious scientific manner precisely because it could neither be isolated nor experimentally experienced. The most sophisticated sociological theories at best give a vague description of a social field as an interlocking network of social and behavioural interactions within specific economic and environmental conditions.

With Maharishi's development of the Technology of the Unified Field, these misconceptions and ambiguities have been removed and the concept of a collective consciousness can and is being tested. The theory states that the collective consciousness of a society is more than the sum total of social interactions; it is a more fundamental reality. The underlying nature of collective consciousness, according to Maharishi, is the field of pure of consciousness, the unified field of natural law.

The field of pure consciousness is a non-localized field of infinite correlation. Thus, "an impulse anywhere is an impulse everywhere." Since the collective consciousness is shared by all, one individual, transcending the surface level of excitation to this universal level of natural law, can generate an influence of coherence and orderliness in the whole society and environment.

For a number of years Maharishi had predicted that when a critical subpopulation of individuals—one percent—experienced and stimulated the underlying field of pure consciousness through the Maharishi Technology of the Unified Field, a type of macroscopic field effect of coherence would occur in the society and the quality of life would improve. This would manifest in more orderly and harmonious individual behavior and a measurable improvement in the various social indices which characterize the quality of life in society.

The first experimental evidence for this phenomenon came in 1974 with the preliminary studies of Borland and Landrith (1976). They observed 11 cities in the United States with populations over 10,000 in which one percent of the population was practicing the TM technique in 1972. They then looked at one major index of the quality of life in these cities—the crime rate—and compared it to the crime rate in 11 other cities matched for population and geographic location. From 1967 to 1972, the crime rate of all the cities had been increasing. The cities with over one percent TM participation showed a mean decrease in crime rate of 8.2 percent from 1972 to 1973, as compared to the year before. The situation in the control cities was significantly different: they showed a mean increase of 8.3 percent during the same year. These findings show that not only is the development of a more ideal society possible, but it requires only a small percentage of the population participating in the TM program. Scientists named this effect the Maharishi Effect, in honor of Maharishi.

Many other more sophisticated studies have confirmed and extended these original findings. An excellent book entitled *The Maharishi Effect* (Oates, 1985) summarizes many of these findings in detail. In the follow sections of this chapter we will highlight the most important studies and consider the possible underlying mechanisms as well as applications of the Maharishi Effect.

RAISING THE COHERENCE OF COLLECTIVE CONSCIOUSNESS—EXPERIMENTAL VERIFICATION

Dillbeck conducted several follow-up studies to Borland and Landrith's initial findings. First he studied the crime rate in suburban cities of metropolitan Kansas City in which there was one percent participation in the TM program for 1976 and 1977. The results of the study showed that the crime rate in the suburbs with one percent TM participation significantly decreased, compared to the suburbs with less than one percent (Dillbeck 1978). Hatchard (1977) found similar results for suburbs of metropolitan Cleveland.

Dillbeck and coworkers (1981) conducted a second,

Improved Quality of City Life: Reduced Crime Rate

Change in Crime Rate 1972–1973: Cities with One Percent of the Population Practicing the TM Technique Compared to Control Cities

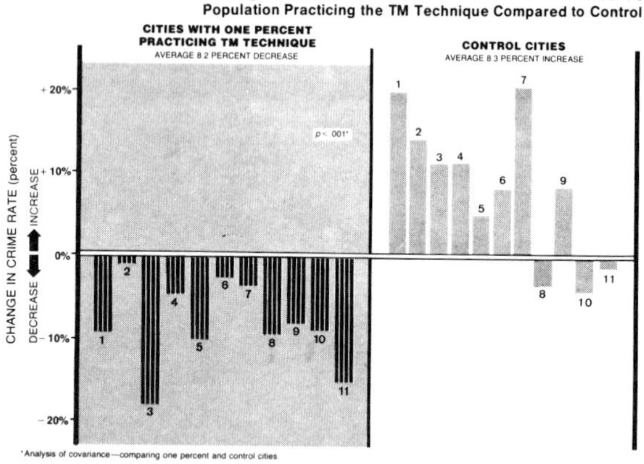

*Analysis of covariance—comparing one percent and control cities.

CITIES WITH ONE PERCENT PRACTICING TM TECHNIQUE

City Name	Percent Change
1. Chapel Hill, NC	− 9.3%
2. Ithaca, NY	− 0.6%
3. Lawrence, KS	− 18.4%
4. Bloomington, IN	− 4.5%
5. Carbondale, IL	− 9.9%
6. Iowa City, IA	− 2.5%
7. Ames, IA	− 3.6%
8. Boulder, CO	− 9.1%
9. Santa Cruz, CA	− 7.9%
10. Santa Barbara, CA	− 8.8%
11. Davis, CA	− 15.2%

CONTROL CITIES

City Name	Percent Change
1. Rocky Mount, NC	+ 20.2%
2. Poughkeepsie, NY	+ 14.4%
3. Lafayette, IN	+ 11.1%
4. Columbia, MO	+ 11.2%
5. Marshalltown, IA	+ 5.0%
6. Oshkosh, WI	+ 8.3%
7. Norman, OK	+ 20.8%
8. Fort Collins, CO	− 3.2%
9. Monterey, CA	+ 8.5%
10. Costa Mesa, CA	− 3.9%
11. Pleasant Hill, CA	− 1.2%

Finding: A retrospective study compared 11 cities in which at least one percent of the population had learned the Transcendental Meditation technique by the end of 1972 to 11 matched control cities with relatively few participants in the TM program, but otherwise comparable in population, location, and crime statistics. Eight of the 11 control cities increased in number of serious crimes from 1972 to 1973 with an average change of+8.3 percent. (For the nation as a whole, the average increase in city crime that year was six percent according to the FBI Uniform Crime Report.) In contrast, the cities with one percent of their population practicing the TM technique decreased in crime rate by an average of 8.2 percent, a relative decrease of 16.5 percent when compared with the control group. This difference in change in crime rate between the two groups of cities was statistically significant ($p<.001$).

Reference: Candace Borland and Garland Landrith III, "Improved Quality of City Life: Decreased Crime Rate," MERU Report 7502, Department of Sociology, Centre for the Study of Higher States of Consciousness, Maharishi European Research University, Weggis, Switzerland, (1975).

more comprehensive, investigation. There were several improvements in this study over the initial research : (1) it included 24 experimental cities, which were all those cities larger than 10,000 population which had one percent TM participation in 1972 (thus no cities were actually excluded) and 24 control cities; (2) an outside consultant was used to help carefully match experimental and control cities; and (3) a number of social and demographic factors were used both to match cities and also to statistically account for the effects. The factors used were population size and density, median years of education, stability of residence, per capita income, percentage of persons in the age range 15 to 29 years, percentage unemployed, and percentage of families with income below the poverty level.

The average data on these factors for the control and experimental cities was quite close. However, the data for crime rate showed that in the five years before the one percent level was reached in 1972, crime rate was actually rising faster in the one percent cities than in the control cities. In 1973, one year after these cities reached the one percent level, there was a marked change in the crime rate trend between these two groups of cities. In the one percent cities there was a sharp average decrease in crime, while in the control cities the crime rate continued to increase. Further, the Maharishi Effect was found to be independent of changes in other social factors.

In a third study, Dillbeck and coworkers (1982) examined 160 U.S. cities over a 14-year period (1964 to 1978) to determine the long-term effects on crime rate in "one percent cities." The study employed a random sample of four groups of 40 cities taken from each of the four largest major population-size groups listed in the FBI Uniform Crime Reports. The study replicated the earlier findings of a relationship between decreased crime rate and TM technique participation, independent of a number of demographic changes. Cross-lagged panel analysis showed that crime rate decrease could be predicted from TM technique participation, while TM participation could not be predicted from crime rate decrease, thus providing evidence that TM participation was the causal factor.

One of the most interesting and unusual studies on the Maharishi Effect was conducted in central Lebanon, in a small village named Baskinta. In a prospective study, the TM program was taught to one percent of the 10,000 people of this village, which was located in a strategic region in the Lebanese conflict. Changes in the experimental village were compared to those in four neighboring villages of similar size and agricultural economic base, over a period of five and one-half years. Two of these village were, like Baskinta, predominantly Christian and under rightist control; two had mixed Christian (native) and Moslem (occupying forces) populations and were under leftist control.

In abrupt contrast to its previous history, there was a complete cessation of hostilities in Baskinta from the time one percent of its population began to practice the TM program, as measured by incoming artillery shells, property damages and casualties. The cessation of violence in Baskinta was in marked contrast to the worsening trends in all the surrounding control villages. Whereas for the previous two years Baskinta has suffered approximately 200 or more incoming shells per season, after it became a "one percent village," the number dropped to zero. However, in the control villages the number of incoming shells actually increased, often exceeding 200 incoming shells per season. Social, economic and ecological conditions also improved in Baskinta as indicated by reported improvements in crop yield, increased social and sporting activities, and accelerated municipal development. The comparative longitudinal design of this study allows exclusion of alternative explanations for these changes, either in terms of seasonal or broader regional changes or in terms of local demographic differences or strategic factors (Abou Nader et al., 1984)

THE SUPER RADIANCE EFFECT

With the introduction of the TM-Sidhi program in 1976, a new and more powerful effect known as "the extended Maharishi Effect," or Super Radiance Effect, began to be studied by several investigators. Maharishi predicted in 1976 that because the TM-Sidhi program represented a more powerful technology for enlivening the unified field, a comparatively small number—the square root of one percent of a population—practicing it in groups should be sufficient to produce measurable effects on the various indices of the quality of life in any society.

The rationale for this prediction came from discussions with leading physicists concerning macroscopic coherent effects seen in physical systems, particularly in the laser. Normal light is emitted in direct proportion to the number of atoms in the light source. When the atoms in a laser are in perfect correlation with each other, then a new type of collective behavior emerges characterized by macroscopic orderliness. The atoms cease to act independently and, instead behave as one complete coherent system. As a result, the intensity of the light emitted by the laser of N atoms is magnified enormously, to N squared. Thus, if 100 atoms are perfectly in phase with one another in the laser, they will radiate with an intensity of 100 squared, or 10,000 times that of a single atom. This phenomenon is known in physics as Super Radiance.

To directly test this prediction, in the summer of 1978 large groups of advanced participants in the TM-Sidhi program were sent to a number of different provinces and states in countries around the world. Their primary activity was to practice the TM-Sidhi program together and raise the coherence of the collective consciousness of that area by spontaneously enlivening the unified field of natural law. In addition, they encouraged government officials and university scientists to test their prediction that the quality of life would be improved merely by the concentrated presence of individuals practicing the TM-Sidhi program.

In the U.S., one of the states chosen was Rhode Island. From June 12 to September 12, 1978, 300 experts went to Rhode Island to demonstrate the effects of the TM-Sidhi program. During that time, approximately 30 to 40 indices of the quality of life improved as compared to the previous summer. Dillbeck and coworkers (1983) reported that murder rate decreased 50 percent, traffic fatalities fell about 48 percent, and suicides decreased 45 percent. Even the weather improved, with 10.3 percent more sunny days. A number of state officials, including those in charge of data evaluation, found these results impressive. Similar results and statistics were reported in other programs around the world.

An even more dramatic program to demonstrate the Super Radiance Effect was conducted in mid-October, 1978, when Maharishi instituted an unprecedented program to create world peace. More than 900 experts in the TM-Sidhi program (representing many nationalities and referred to by Maharishi as Governors of the Age of Enlightenment) were sent for several months to troubled areas of the world (Central America, Southeast Asia, the Middle East, and southern Africa). In most cases, there was a remarkable time-correlated decrease in violence and disorder in each of the trouble spot countries shortly after the arrival of the TM-Sidhi participants, significant progress towards peace while they remained, and a rebound to increased violence upon their departure.

Orme-Johnson and coworkers (1979) analyzed the effects of this World Peace Project. They examined the number of deaths due to the war in Zimbabwe, including terrorist collaborators, civilians and security forces before, during, and after the arrival of the TM-Sidhi participants. Before their arrival (on November 4) the number of deaths was 15.8 persons per day. On the arrival of 56 TM-Sidhi participants, the violence was immediately reduced to zero deaths, and the death rate remained low (3.8 persons per day) for the next 11 days. When the TM-Sidhi participants split into two groups of 20 and 36 persons and went to different cities, the death rate approximately doubled. As Dr. Orme-Johnson suggests,

> An explanation for this increase is provided by the physics of coherent states in which the effect of a coherent system is proportional to N^2 whereas the effect of an incoherent system is proportional to N. With the TM-Sidhi practitioners as a coherent system, all in one group, the N squared effect equals 56 squared or 3136. With the Governors in two groups the effect equals 20 squared plus 36 squared or 1696, 46% less power than the same number of people in one group. Thus, splitting into two groups cut the predicted effectiveness by half and the death rate due to the war was seen to double. When the Governors reduced their TM-Sidhi program in preparation to return to the U.S.A., on November 28, the death rate immediately increased.

A further independent confirmation of these findings comes from evidence provided by the Conflict and Peace Data Bank (COPDAB), the largest data bank on international conflict and resolutions. The COPDAB file, collected for over thirty years at the Center of International Development at the University of Maryland, uses a procedure known as content analysis, in which two independent raters assess publications and assign news from a wide variety of sources into areas such as hostile acts or

Increased International Harmony

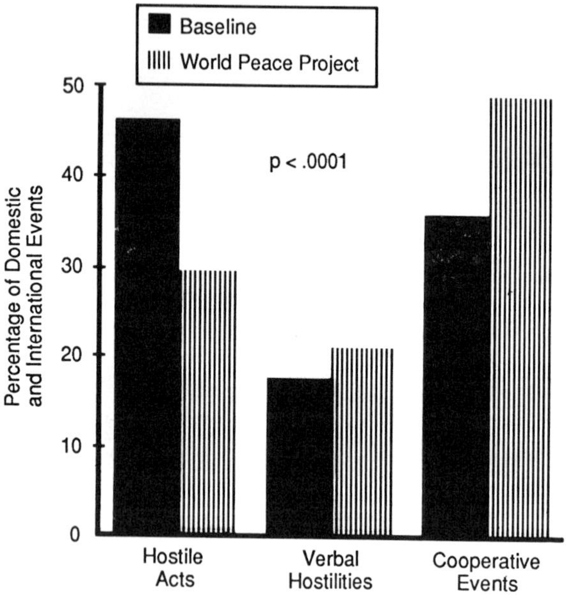

During a ten-week World Peace Project (October 8 - December 23, 1978) a total of 1,400 experts in the Transcendental Meditation and TM-Sidhi techniques went in groups to the world's five major trouble-spot areas in order to test the hypothesis that the group practice of this technology would create coherence in a field of collective consciousness, thereby restoring balance to the socio-political systems in those areas and in the world.

Evaluation of the project in the trouble-spot countries using a large independent data source on international conflict, the Conflict and Peace Data Bank (COPDAB) file, indicated that there was a significant proportional decrease of Hostile Acts and an increase of Cooperative Events in domestic affairs and international relations during this period compared with the preceding ten-week baseline period (chi-square = 18.51, $p<.0001$). By using the prior ten years as a control, this analysis showed that the results could not be attributed to weekly, monthly, or yearly cycles or to variability in the size of the weekly data sets.

Reference: David W. Orme-Johnson, Charles N. Alexander, and Jean G. Bousquet, "Impact Assessment Analysis of the Effects of Coherence Creating Groups on International Conflicts," (Department of Psychology, Maharishi International University, USA, 1985) *Scientific Research on the Transcendental Meditation and TM-Sidhi Program: Collected Papers*, Vol. 4, in press, (MIU Press, 1985).

cooperative events, based on a predetermined scale.

The COPDAB file showed that when the ten weeks before the World Peace Project were compared to the ten weeks during the project, there was a significant decrease in hostile acts in the trouble-spot areas, from nearly half of all events (46.6 percent) to less than one third (29.5 percent), and a significant increase in cooperative acts, from 36 percent to 49.2 percent. When the ten-week period of the Peace Project was compared to the same period during the previous ten years (1968-77) it was again shown to be markedly more peaceful (Orme-Johnson et al., 1985).

The results of one of the intervention studies assessing the effectiveness of the Super Radiance Effect were particularly striking. This was the International Peace Project conducted in Israel in the late summer of 1983. The project was initiated by Dr. Charles Alexander, then research faculty at Harvard University and Director of The Institute for Research on Consciousness and Human Development. The study was supported in part through a grant from the Fund for Higher Education in New York and conducted in cooperation with Orme-Johnson and other coworkers (Orme-Johnson et al., 1984a).

This was a prospective study in which certain results were predicted before its actual implementation. The study aimed to gather in Jerusalem 200 experts in the Maharishi Technology of the Unified Field, the square root of one percent of the population of Israel, to collectively practice the TM and TM-Sidhi program. Because many of the participants had professional responsibilities, the actual number of participants varied from week to week, reaching 200 during one two-week period and several weekends. This constantly varying pattern actually enabled a more careful statistical analysis of the Super Radiance Effect.

The results of the study clearly demonstrated the

effect: when Super Radiance numbers went up, there was a significant decrease in reported war deaths in Lebanon, the intensity of the Lebanon war (as measured by news content analysis), automobile accidents and fires, and an increase in the Israeli stock market. Using a statistical procedure known as time series analysis, a clear relationship was determined between the overall quality of life in both Israel and Jerusalem (as measured by a composite of different variables in a "quality of life index") and the number of Super Radiance participants.

THE GLOBAL MAHARISHI EFFECT—A TASTE OF UTOPIA

One of the most ambitious and widely publicized demonstrations of the Super Radiance Effect took place in Fairfield, Iowa, home of Maharishi International University, at the end of 1983. This demonstration, called "A Taste of Utopia," was announced as a "global sociological experiment" and involved more than 7000 people—the square root of one percent of the world's population—collectively practicing the TM-Sidhi program. Prior to the assembly, a number of predictions were made, including an improvement in international relations and in the major world stock markets, and a worldwide decrease in crime, accidents and illness. In each case these predictions were confirmed (Orme-Johnson et al., 1984b. The effects were so widespread that the entire phenomenon was called the Global Maharishi Effect.

For example, the World Stock Index, a single measure of stock prices (the weighted average of the 19 most important stock markets in the world), had been generally trending down for three weeks prior to the course. At the onset of the course the index began to rise, and rose steadily until the first Monday after the assembly ended, when it suddenly declined again. Eighteen of the nineteen markets included in the World Stock Index increased, eight of the eleven largest markets in the world broke all-time records, and the United States market, after a long downward trend, suddenly rallied during the three weeks of the Assembly. During the same three-week period in the previous five years, an average of one-half of the markets went up and one-half went down.

In other areas of the study, traffic fatalities over the 1983 Christmas and New Year holiday dropped to an all-time low, even though miles driven were at an all-time high. There were 31 percent fewer fatalities per mile driven than in the previous 16 years. Infectious diseases dropped by 15 percent compared to the three-week periods before and after the assembly, and by 32 percent compared to the mean of the same time period during five previous years. Patent applications increased 15 percent over the amount normally predicted by the United States Patent Office, and then decreased to normal values immediately after the assembly (these statistics take into account seasonal increases).

Regarding international affairs, content analysis of newspapers revealed that for the three weeks before the assembly, positive events comprised 18 percent of the total number of events worldwide. This percentage increased to 31 percent during the assembly, and then decreased to 13 percent afterward.

One of the intentions of the assembly was to decrease violence in one of the most troubled areas of the world, Lebanon. Actually, in addition to the Taste of Utopia assembly, two other assemblies were later held as part of a long-term study on the effects of the Super Radiance program on violence in Lebanon. The second assembly was held in March, 1984, within Lebanon itself and involved some 60 experts in the Maharishi Technology of the Unified Field, a number large enough to create the Super Radiance Effect for Beirut and surrounding areas. The third assembly was a group of 2000 experts held in Yugoslavia in late April and early May, 1984.

NUMBER OF EXPERTS IN THE MAHARISHI TECHNOLOGY OF THE UNIFIED FIELD

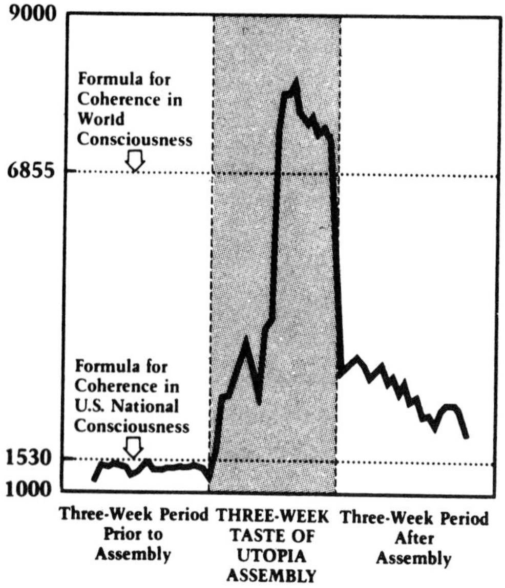

SOURCE: Capital of the Age of Enlightenment

ORME-JOHNSON, D. W.; CAVANAUGH, K. L.; ALEXANDER, C. N.; GELDERLOOS, P.; DILLBECK, M.; LANFORD, A. G.; and ABOU NADER, T. M. The influence of the Maharishi Technology of the Unified Field on world events and global social indicators: The effects of the Taste of Utopia Assembly. Department of Psychology and Department of Management and Public Affairs, Maharishi International University, Fairfield, Iowa, U.S.A., and Department of Nutrition and Food Service, Massachusetts Institute of Technology, Cambridge, Massachusetts, U.S.A., 1984.

RISING WORLD STOCK INDEX

World Index of Capital International S.A., Geneva

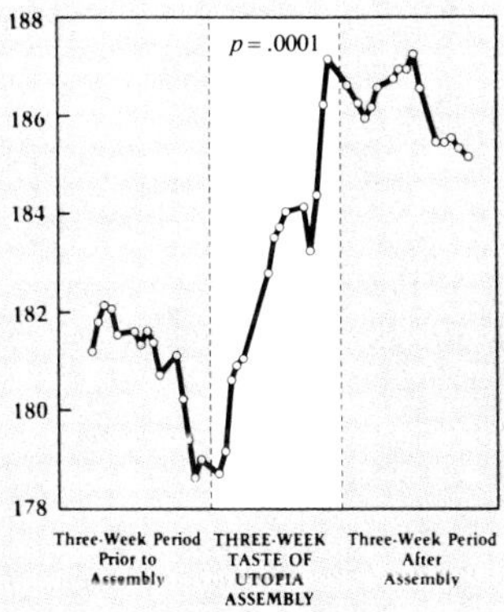

SOURCE: *The Wall Street Journal*

During the Taste of Utopia Assembly rising coherence from MIU created a wave of confidence and optimism throughout the world, seen as a rise in the World Stock Index, a composite of 19 countries. *After the Assembly the Index resumed a downward trend.*

INCREASED PATENT APPLICATIONS

Percent Change in the Actual Number of Patent Applications Compared to Predicted Number

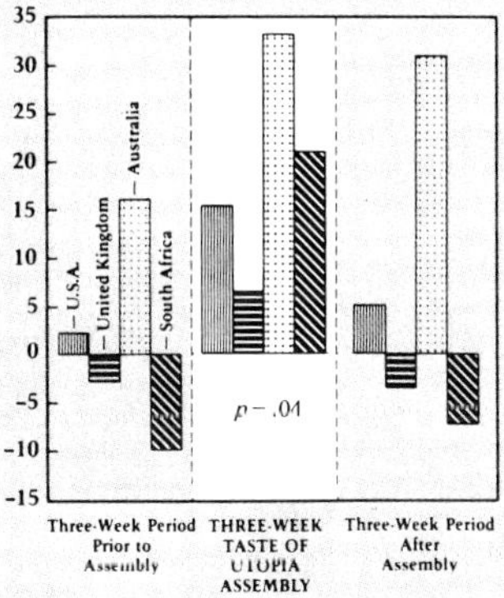

SOURCES: U.S. Patent Office; U.K. Patent Office; Australia Patent, Trade Marks, and Design Office; Government Patent Office, South Africa

During the Taste of Utopia Assembly the number of patents filed, an important measure of national creativity, increased significantly in countries on four continents. *After the Assembly the number of applications tended to revert to a pattern similar to that seen prior to the Assembly.*

SIMULTANEOUS INCREASE OF MAJOR STOCK MARKET INDICES

Percent Change over Three-Week Period

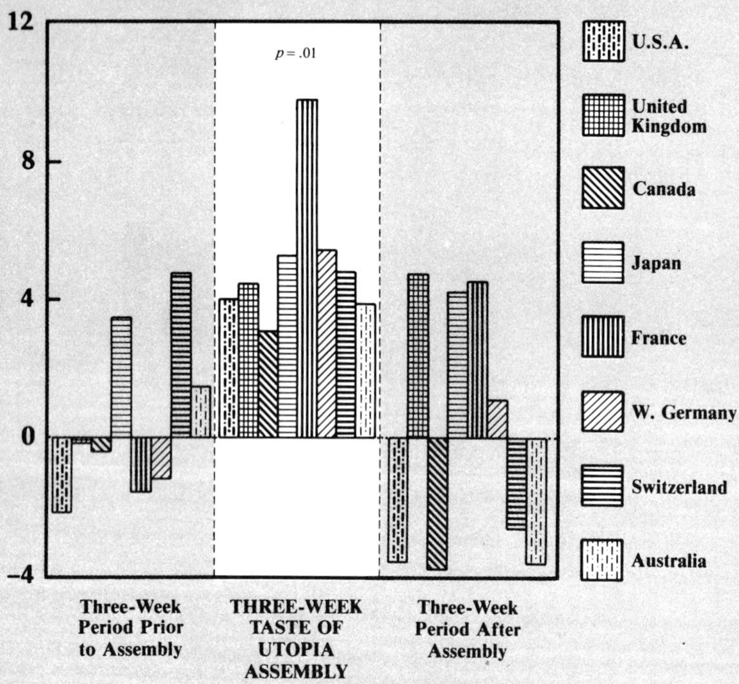

SOURCE: Capital International S.A., Geneva
(The Wall Street Journal)

During the Taste of Utopia Assembly the major stock markets of the world increased simultaneously, indicating balanced economic growth worldwide. *After the Assembly the same major stock markets reverted to a pattern similar to that seen prior to the Assembly, with some increasing and some decreasing.*

DECREASED TRAFFIC FATALITIES

Percent Decrease in Traffic Fatalities, 17 December 1983 to 6 January 1984, Compared with the Number Predicted from Prior Years

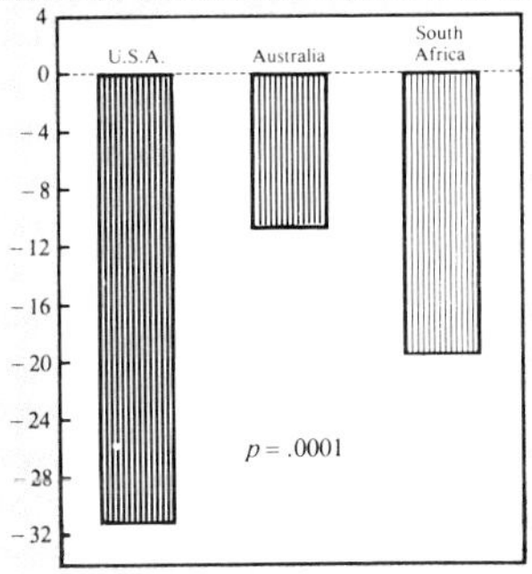

SOURCES: National Safety Council, U.S.A.; National Road Safety Council, South Africa; State Police Dept., Sydney; Police Dept., Perth, and Road Traffic Authority, Hawthorn, Australia.
(Note: Traffic fatalities for Australia are for Western Australia, New South Wales, and Victoria.)

Traffic fatalities decreased significantly during the Taste of Utopia Assembly. In the U.S.A., traffic fatalities per day over the Christmas and New Year's weekends were at an all-time low, even though miles driven per day, despite the cold weather, were at an all-time high.

WORLDWIDE DECREASE IN AIR TRAFFIC FATALITIES

Number of Air Traffic Fatalities From 17 December to 6 January

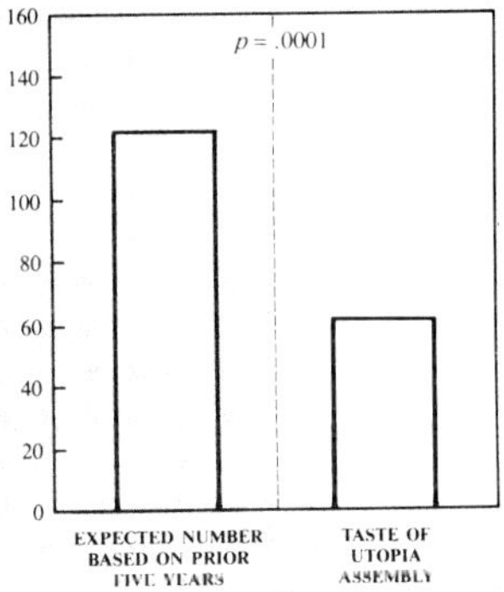

SOURCES: International Civil Aviation Organization; National Transportation Safety Board, U.S.A.

During the Taste of Utopia Assembly the number of air traffic fatalities in the world was 49% lower than the expected number based on the prior five years for the same time of year. It was also 29% lower than the lowest number during the equivalent three-week period in the prior five years.

DECREASED INCIDENCE OF INFECTIOUS DISEASES

Percent Change in Total Incidence of Notifiable Infectious Diseases Compared to Medians for Equivalent Periods in Previous Years

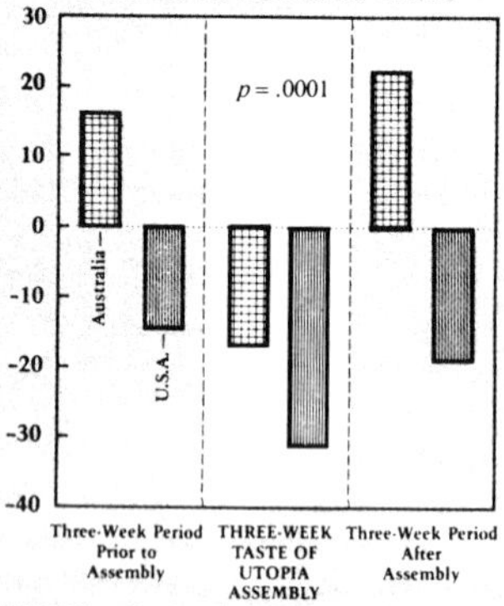

SOURCES: Center for Disease Control, U.S.A.; Department of Health, Commonwealth of Australia. (Note: Number of previous years considered—4 years for Australia; 5 years for U.S.A.)

The incidence of all major categories of notifiable infectious diseases decreased during the Taste of Utopia Assembly. *After the Taste of Utopia Assembly the incidence of infectious diseases began to rise again towards levels comparable to previous years.*

REDUCED CRIME TOTALS

Percent Change in Daily or Weekly Crime Totals

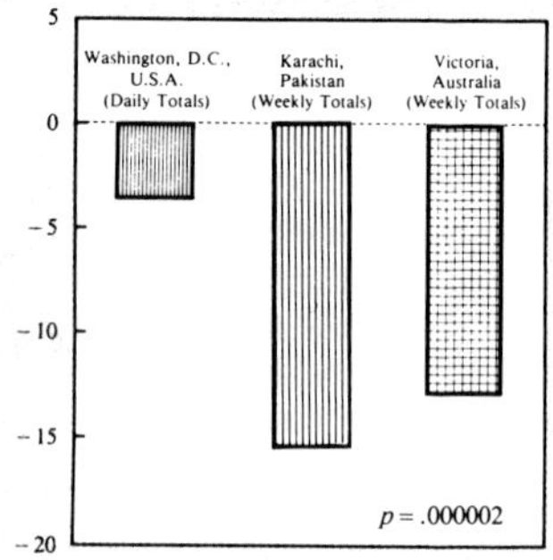

SOURCES: City Police Department, Washington, D.C.; Inspector-General of Police, Government of Sind, Karachi, Islamic Republic of Pakistan; Research and Development Department, Victoria Police, Melbourne, Australia.

Using time series analysis it was found that during the Taste of Utopia Assembly significant decreases occurred in daily or weekly crime totals, in locations on three continents, in comparison with the average daily or weekly totals for the 24 weeks prior to and three weeks after the Assembly.

INCREASED POSITIVITY OF EVENTS IN SITUATIONS OF INTERNATIONAL CONFLICT

Percent of Total Events as Rated for Degree of Conflict

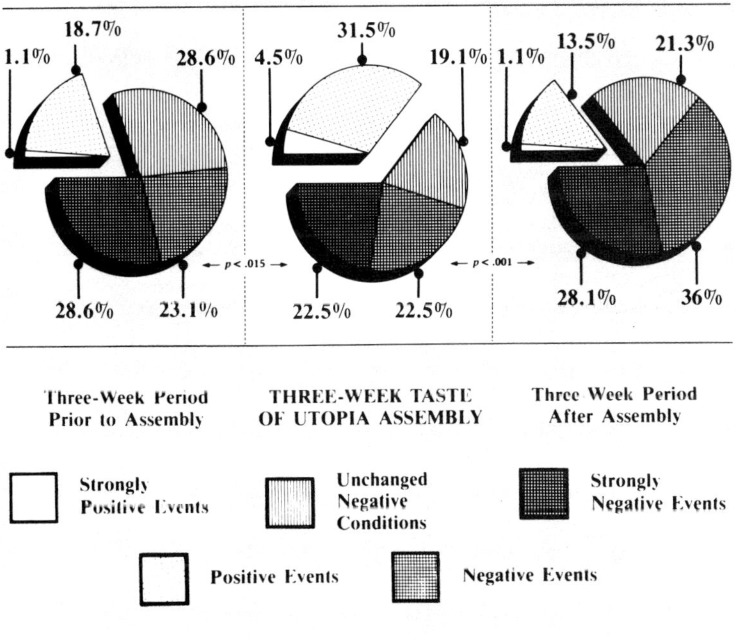

SOURCE: Content Analysis of *The New York Times*

During the three-week period of the Taste of Utopia Assembly the balance of negativity to positivity in events pertaining to international conflicts in the trouble-spot areas of the world shifted significantly towards increased positivity. *After the Assembly the balance of events reverted towards increased negativity.*

Reduction of the War in Lebanon — 1984

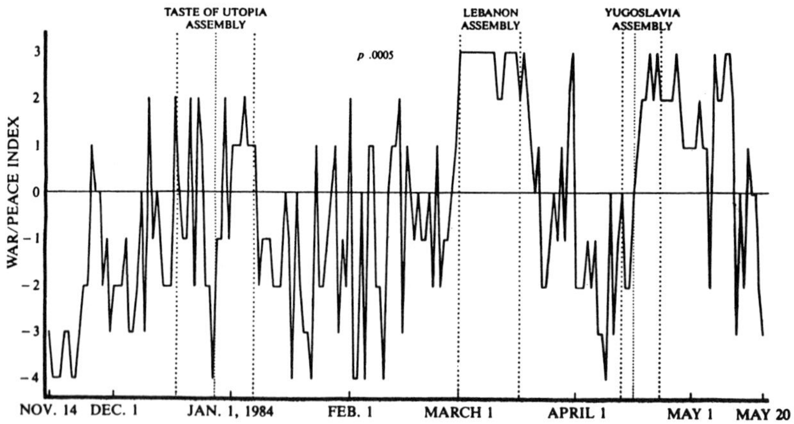

For the six-month period from November 13, 1983, to May 18, 1984, the daily level of the Lebanon Peace/War Index fluctuated widely and was generally negative. In contrast, during the three creating coherence assemblies held at this time (while the number of participants exceeded the predicted threshold for an influence on the war) there were plateau periods of progress toward peaceful resolution of this conflict. Time series analysis indicates significantly greater progress toward peace during these assemblies than would have been expected from the prior history of the Lebanon war ($p = .000046$). (Data source: daily Beirut newspapers, Al Nahar, Al Anwar, Le Reveil, and L'Orient.)

Reference: C.N. Alexander, T.M. Abou Nader, K.L. Cavanaugh, J.L. Davis, M.C. Dillbeck, R.J. Kfoury, D.W. Orme-Johnson, "Effects of the Maharishi Technology of the Unified Field on the Conflict in Lebanon: A Time Series Analysis of the Influence of International and National Creating Coherence Assemblies," *Scientific Research on the Transcendental Meditation and TM-Sidhi Program: Collected Papers*, Vol. 4, in press, (MIU Press, 1985).

Three separate parameters were utilized to objectively measure the effects of these assemblies. The first two were war-related injuries and war-related deaths (public statistics gathered from the Lebanese government), and the third was a peak/war index compiled through content analysis of Lebanese newspapers, using raters from the three dominant factions, Christian, Moslem and Druse.

The study extended from one month before the Taste of Utopia assembly to one month after the Yugoslovia course ended—nearly 250 days in all. Using time series analysis the researchers found that there was a significant average drop—55 percent—in war deaths per day during the three Super Radiance assemblies. Content analysis yielded a similar picture of a sudden jump in positivity during the assemblies. (Alexander et al., 1984).

More than 20 studies have been conducted on the Super Radiance Effect to date. In addition to studies already mentioned, several studies have looked at other factors such as traffic accidents and fatalities (Landrith and Dillbeck, 1983; Burgmans et al., 1982; Dillbeck et al., 1984) weather patterns (Rabinoff, et al., 1981) economic indices (Lanford, 1984a; Beresford and Clements, 1983; Cavanaugh et al., 1984) and homicides (Lanford, 1984b). Several further studies have been conducted at other large world peace assemblies held in the U.S. and around the world (Davies and Alexander, 1983; Dillbeck et al., 1983; Dillbeck et al., 1984b).

In July of 1985, another large assembly of some 5600 experts in the TM-Sidhi program was held in Washington, D.C.. It was predicted that since this assembly was held in the nation's capital, less than the required number of 7000 would be need to produce global effects. Time series analysis showed that, as in the Taste of Utopia assembly, there were a number of significant improvements: the Dow Jones Industrial Average increased significantly, the World Index of international stock prices (19 countries) increased, international conflicts decreased, infections diseases decreased, fires in Washington, D.C., decreased and U.S. patent applications increased (Orme-Johnson and Dillbeck, 1985).

One study was especially convincing. Twelve indices of the quality of life in the U.S. were monitored over a 23-year period from 1960 to 1983 (Orme-Johnson and Gelderloos, 1984). An overall Quality of Life Index was formulated using the following: (1) crime rate, (2) percentage of civil cases reaching trial, (3) infectious disease rate, (4) infant mortality, (5) suicide rate, (6) cigarette consumption per capita, (7) alcohol consumption per capita, (8) gross national product (GNP) per capita, (9) patent applications, (10) number of educational degrees conferred, (11) divorce rate, and (12) traffic fatalities. The trend of the Quality of Life Index is clearly downward over the first part of the 23-year period, worsening rapidly after 1967 until 1975, when it reached its lowest point. It improved slightly over the next six years, until 1982, when there was a sudden increase in the rate of improvement.

In 1975 and 1976, the first turning point in the trend, the number of people learning the TM technique significantly increased. More than a quarter of million people in the U.S. learned TM in 1975. Interestingly, Maharishi referred to 1975 as "the Dawn of the Age of Enlightenment" even though this was a time of economic and social turbulence. One year later, Maharishi introduced the TM-Sidhi program and began emphasizing the importance of gathering a sufficient number of people together to use this more advanced procedure for improving the collective health of the world. It took almost seven years—until 1982—for the community of experts in the Maharishi Technology of the Unified Field in Fairfield, Iowa, to grow to the required number of about 1,600 (the square root of one percent of the population of the U.S.). Since 1982, this group has collectively practiced the TM-Sidhi program in Fairfield, and since that time there has been a marked

Improved Quality of Life

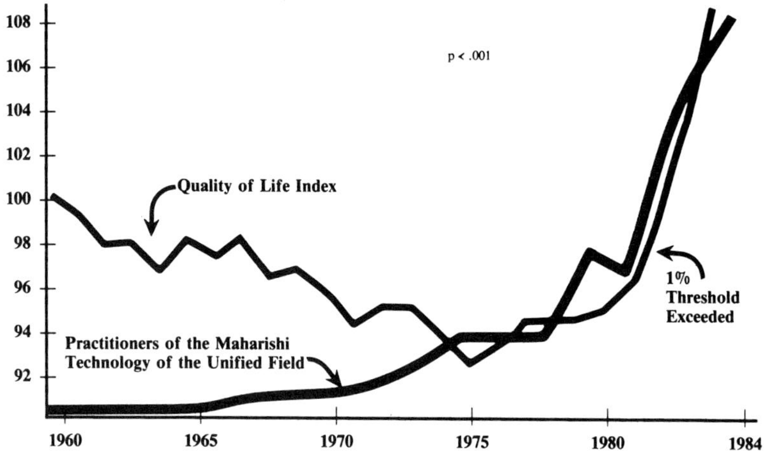

The quality of life in the United States as measured by an index comprised of 12 variables (crime rate, percentage of civil cases reaching trial, infectious diseases rate, infant mortality rate, cigarette and alcohol consumption rate, GNP per capita, patent application rate, degrees conferred rate, divorce rate, traffic fatalities rate, and hospital admissions rate) was found to decrease almost constantly from 1960 to 1975. In 1975 and 1976 several 100,000's of people were instructed in the Transcendental Meditation technique, bringing the rate over .4% in 1976. Cross correlation showed that the rate of practitioners of the TM technique had a leading relationship on the development of the quality of life. Since 1976, the quality of life has been improving continously, and the improvement started to accelerate sharply, once the 1% threshold — based on the combined effect of the number of meditators and the people engaged in the collective practice of the TM-Sidhi program at MIU, Fairfield — was exceeded in 1982, 1983 and 1984. The rate of improvement in Iowa, MIU's home state, has been significantly higher in these years.

The highly significant correlations ($p < .0001$) between the development of the quality of life and the proportion of people engaged in the Transcendental Meditation and TM-Sidhi program suggest that the practice of the Maharishi Technology of the Unified Field induces coherence throughout all levels of society, leading to improvements of all aspects of life. The temporal and spatial trends support this causal relationship strongly. Predictions on basis of the regression model indicate that the quality of life would improve tenfold every 4 years once a group of 7000 experts of the Maharishi Technology of the Unified Field would be established in the United States.

Reference: David W. Orme-Johnson and Paul Gelderloos, "The Long-Term Effects of the Maharishi Technology of the Unified Field on the Quality of Life in the United States (1960-1983). Updated version. (Department of Psychology, Maharishi International University, USA 1984). *Scientific Research on the Transcendental Meditation and TM-Sidhi Program: Collected Papers*, Vol. 4, in press, (MIU Press, 1985).

improvement in almost all indices of the quality of life in the whole nation.

A further addition to this study is an examination of the quality of life specifically in Iowa, where the Super Radiance Effect would be expected to be greatest. Of the twelve variables used in the Quality of Life Index, six are available in Iowa (crime, infectious disease, infant mortality, alcohol consumption, divorce and traffic fatalities). Since 1982 Iowa has fared better on all of these than the rest of the nation.

THE MECHANISM OF THE MAHARISHI EFFECT

Because of its extremely long-range and subtle effects, a theoretical understanding of the Super Radiance Effect must ultimately rely upon unified field theories. Some preliminary research on EEG coherence, however, has revealed very interesting data which may be useful for future models. The experimental hypothesis of this research was that the field effects of consciousness stimulated by the group participation in the TM-Sidhi program could affect the EEG coherence between subjects even at a large distance. EEG coherence was chosen since it increases during TM.

The design of the study was as follows: three subjects were assigned to three separate sound attenuating rooms to practice their usual TM program for one hour. Four frontal and central EEG channels were measured on each subject for six experimental days and six control days. On the six experimental days the EEG coherence measurements on the three meditating subjects were taken during a 15-minute time-period in which a group of 2500 individuals 1000 miles away were practicing the TM-Sidhi program. The three EEG subjects did not know the exact time of this 15-minute period. In fact, the three began their meditation period at different times—15, 20, or 25 minutes before the experimental period—and the starting times were randomized, with the constraint that each starting time was used twice by each subject. Inter-subject coherence was computed using a novel procedure in which researchers compared the EEG from the same lead on pairs of subjects. As a control procedure, the same subjects were measured at the same time of day under identical laboratory conditions for six additional control days after the course was over and the group of 2500 individuals had dispersed.

Inter-subject coherence during the experimental period was compared with inter-subject coherence during the preceding 15-minute period, which was used as a baseline. The change in coherence from the baseline to the experimental period on experimental days was compared with changes during equivalent time periods on control days.

The mean inter-subject coherence tended to be low, and varied in magnitude in different frequency bands. For example, alpha and beta coherence were approximately 5 standard errors of the mean higher than delta coherence on both experimental and control days. More importantly, it can be seen that coherence tended to increase by approximately two standard errors during the experimental period compared with the baseline on experimental days and that the increase was less on control days. This effect was particularly evident for the alpha and beta frequencies.

Multivariate analysis of variance showed a significant interaction of periods (baseline and experimental) and days (experimental and control). This interaction indicated that inter-subject coherence over all frequencies and derivations increased significantly more during the experimental period relative to the baseline period on experimental days than on control days, which was consistent with the hypothesis of the experiment.

Thus these results, although still preliminary, provide data suggesting that the field effects of consciousness can be physiologically measured. While we still do not understand the mechanism by which these field effects occur, considerable animal research has shown that many species have extremely responsive sensory systems (sensitive, for example, to very weak magnetic fields), which enable them to navigate over large distances (Beischer, 1971). It has also been shown that the human eye is responsive to one or two photons of green light (Bouman, 1961). Adey and coworkers have conducted a considerable amount of research which has shown that calcium efflux can be induced across brain tissue under the influence of extremely weak electromagnetic fields (Orme-Johnson et al., 1982).

These and other studies have lead some researchers to suggest that, in addition to the already known classical mechanisms of neural transmission, there may be other modes of functioning which use far more sensitive mechanisms than those previously conceived—perhaps even quantum mechanical modes of functioning. We will discuss these ideas in greater depth in the next chapter, when we consider the relationship of this field effect of consciousness to the laws of nature described by modern physics. Whatever the precise theoretical explanation of these results and those of previous studies on the Maharishi Effect, they do indicate a remarkable new technology for improving the quality of life and establishing harmony and peace in different countries throughout the world.

THE PHYSIOLOGY OF WORLD PEACE

Like a living organism, the health of a society depends upon the health of its internal parts. To function properly it must have a high level of internal stability, integration among systems, efficient self-purification devices, and the capacity for creative expression and growth. Society can, in effect, be considered to have its own physiology. As in the individual, the longevity of the society and its cultural values is dependent upon the self-organizing power of its physiology, that is, how capable it is of maintaining a state of low internal entropy in the midst of high external change.

Many nations today, especially developing ones, face a very difficult problem. They need and want to maintain their cultural identity, yet they also need and want to incorporate the technology and comforts enjoyed by developed nations. There is often a conflict between the values of their cultural heritage and the values of technologically advanced countries. Unfortunately, the cultural heritage, the very root and strength of the nation, is often sacrificed in favor of the need for technological growth.

How can these countries preserve the precious values of their traditions and at the same time progress? Essentially the same problem faces any living system—how does it preserve its internal order amidst change? It is necessary to enliven an internal source of order within, to develop deep within the system such collective coherence that the inner structure remains fluid yet stable, able to adapt to change by both incorporating beneficial and supportive elements and at the same time resisting or eliminating those that are harmful.

According to Maharishi, this internal source of order within society is precisely the same as that within each individual—pure consciousness. Maharishi states that the loss of knowledge of this internal source of order as the prime mover of life and the organizing power of society is the reason for the many social and international problems we see today.

The key to ideal social behavior, as Maharishi explains, is action in accord with the laws of nature. The individual's awareness must be established in pure consciousness, the home of all the laws of nature, so that automatically and spontaneously his actions are right—in tune with the needs of the environment and producing minimum disorders. When enough people are acting from this most refined level of consciousness, the collective consciousness of the society or nation in which they are living becomes more coherent and is brought in tune with the laws of nature. As Maharishi says:

> The real governor of the nation is its collective consciousness. National consciousness is the sum total of individual consciousness. So national consciousness can be easily handled from the subtle, quiet, holistic value of the individual's consciousness. Invincibility of the nation means perfect health of the nation. The Transcendental Meditation and TM-Sidhi program provides a very simple, natural, and effortless formula to handle that value which is fullness in life. The result is perfect health in the individual and invincibility for the nation.

In order to make the knowledge and technology of consciousness available to all nations, Maharishi, on January 12, 1976, founded the World Government for the Age of Enlightenment. In his address inaugurating the World Government of the Age of Enlightenment, Maharishi explained its role.

> The role of the World Government of the Age of Enlightenment is parental, because whatever political system a government may wish to adopt, the World Government wishes only to give each government the ability to succeed in fulfilling its aspirations. The fights between different political systems—communism, capitalism, democracy, dictatorship—have

> occurred because the knowledge of the common feature that binds together all life everywhere, the field of consciousness, has not been available. It is variety that makes a world, and it is variety that decorates the world. Different systems of political organization and social order are like the different colored flowers in a flower arrangement. With all its variety of colors the arrangement becomes beautiful. The world was designed to have variety. It is not the difference between political systems that is at fault. What is at fault is the lack of knowledge of the common field of nourishment that gives life to every system everywhere. It is obvious that no system of government is perfect, because no system has produced an ideal field of consciousness. The World Government of the Age of Enlightenment is going to strengthen all systems of government and bring them all to function under the direct evolutionary influence of nature.

The World Government is a non-political, non-religious global organization with the participation of more than 100 countries. The World Government of the Age of Enlightenment has repeatedly offered its readiness to enter into contractual agreements to solve the problems of any government regardless of the magnitude and nature of the problem—political, economic, social or religious—and irrespective of its system—capitalism, communism, socialism, democracy, or dictatorship.

Maharishi explains that the sovereignty of the World Government is in the domain of consciousness—the prime mover of life, the ground state of natural law, the field of all possibilities. The fundamental principle underlying the programs of the World Government of the Age of

Enlightenment is the Vedic realization that knowledge has organizing power.

Unmanifest pure knowledge is the hidden organizer and administrator of all phenomenal creation. Therefore, a government that succeeds in harnessing the organizing abilities of the field of pure knowledge, the unified field of natural law, will experience success in internal and external affairs. As can be seen from the extensive research on individual and social effects of the Maharishi Technology of the Unified Field, the application of this technology promises for the first time in world history to bring to every nation integrated national consciousness, cultural integrity, and self-sufficiency; and to the whole family of nations, world peace in a unified field based civilization.

Chapter 9

NEUROPHYSIOLOGICAL MODELS OF CONSCIOUSNESS

There is a compelling desire among neuroscience researchers today to understand the nature of human behavior and human consciousness in terms of the structure and function of the parts of the brain. Innumerable theoretical models have been constructed at various levels of organization within the brain to explain such phenomena as memory, learning, motivation, and mental health. As mentioned in the first chapter, the present approach of neuroscience is similar to the classical approach of physics. All the attention is on the more manifest level of localized excited states or parts; very little attention has been placed, and very little progress has taken place, in understanding the lesser excited and more integrated states of brain functioning.

The Maharishi Technology of the Unified Field provides modern neuroscience with a new research tool for understanding the brain in its simplest, least excited, least localized and most integrated state—the state of pure consciousness. Based on the physiological and sociological research on the TM and TM-Sidhi program a number of models of brain functioning during the state of pure consciousness have been developed. In this chapter we will discuss these proposed models in detail as well as future directions of research, which these models suggest and which might help further elucidate the mechanisms they propose.

These models can be classified according to their degree of sophistication and their ability to explain the experimental findings concerning the nature of pure consciousness.

We can consider several levels of theoretical models: Level 1 corresponds to earlier models, which are primarily concerned with explaining the physiological changes during the TM technique and characterizing, in terms of a classical reductionistic model, the neurophysiological mechanism underlying the state of pure consciousness and the growth to higher states of consciousness. Level 2 corresponds to more advanced models, which go beyond the classical reductionistic approach and attempt to explain the more extended field effects of consciousness in terms of recent discoveries in quantum field theory.

While the Level 1 models are useful in describing the neural mechanisms that support the state of pure consciousness during the TM and TM Sidhi program, they nevertheless remain at a superficial level. Only the Level 2 models offer an entirely new and profound approach to comprehending the neurophysiological basis of higher states of consciousness.

THE NEUROPHYSIOLOGICAL BASIS OF STATES OF CONSCIOUSNESS—LEVEL 1

What are the neurophysiological mechanisms responsible for the state of pure consciousness? In order to discuss them it will be helpful to consider the models for waking, dreaming and sleeping states that have already been proposed. In these models a number of important major brain structures have been implicated, such as the thalamus, reticular formation, brain stem nuclei, limbic system, and pre-frontal cortex. Let us consider each of these structures briefly.

The thalamus is located in the center of the brain and acts as a relay station for sensory information as it travels from a sense organ to the cerebral cortex. Cells within the thalamus project to sensory and association areas of the cortex; thus, their synchronization may in turn cause widespread synchronization within the cortex. Certain systems or centers within the thalamus have been suggested as being responsible for the maintenance of major states of consciousness. The diffuse thalamus system, a system of cells that indirectly has widespread connections throughout the cortex, has been proposed to cause internal inhibition leading to sleep.

The reticular formation, and more specifically the reticular activating system (RAS), has been proposed to cause internal excitation leading to wakefulness (Magoun, 1963). Anatomically, the RAS consists of a number of clusters of cells, which are spread throughout the brain stem. Some researchers have identified specific groups of cells within the RAS (known as the nucleus reticularis pontis oralis) as being the most powerful activators of EEG desynchronization and alert wakefulness.

Slow wave sleep seems to occur both when there is a decrease in the input activity of the RAS and sensory pathways in the diffuse thalamic systems and when there is an activation of a specific group of cells within the brain stem known as the raphe system. The cells of the raphe system have been shown to contain a high concentration of the neurotransmitter serotonin. Activating the raphe center inhibits the arousing action of the reticular activity center, allowing the diffuse thalamic system to drive the cortex into the EEG patterns characteristic of sleep.

Dream sleep is believed to be activated and maintained by at least two other centers within the brain stem, the locus caeruleus and the giant reticular neurons. Activation of the locus caeruleus neuron via the release of another neurotransmitter, norepinephrine, is believed to result in the phenomenon of slow wave sleep (Jouvet, 1967).

Other major structures such as the frontal lobes of the

Brain Structures Maintaining Major States of Consciousness

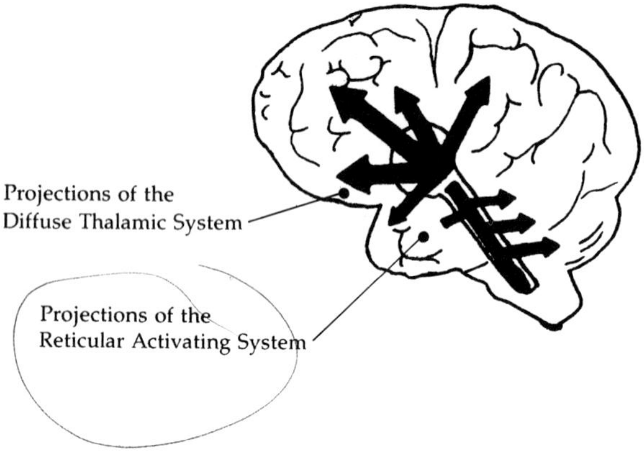

Reference: Magoun H.W. (1963). *The waking brain.* Springfield, Ill., Charles C. Thomas.

Brain Centers Proposed for Waking, Sleeping and Dreaming

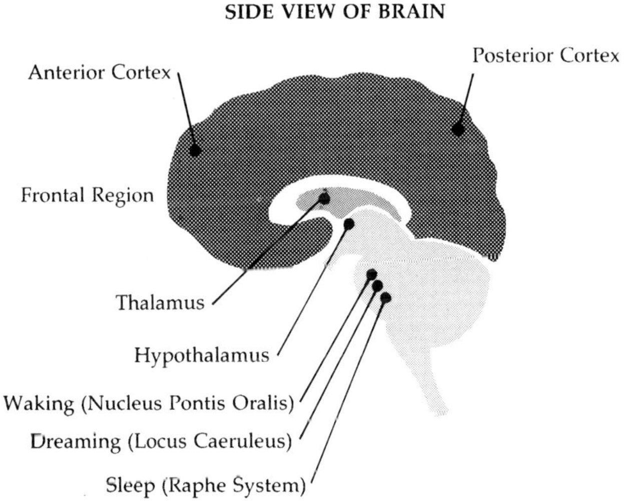

Reference: Jouvet M. (1967). The states of sleep. *Scientific American* 216: 62-72.

cerebral cortex and limbic system are also believed to be important in influencing EEG activation by modulating the reticular activating centers within the brain stem and affecting other physiological functions. For example, one of the structures within the limbic system, the hypothalamus, has been postulated to be involved in the regulation of sleep and wakefulness. The hypothalamus is also a major integrative center in the brain responsible for such functions as heat regulation, water balance, food intake, and endocrine and autonomic functioning.

Gellhorn (1967) emphasizes the influence of the hypothalamus and its control of autonomic and somatic activity in all states of consciousness. He suggests that the hypothalamus and RAS are the primary structures whose excitation and inhibition determine the particular state of consciousness. Instead of localizing the control of various states of consciousness to only specific areas within the RAS, Gellhorn emphasizes the importance of integrated action among the anterior and posterior regions of the hypothalamus, certain parts of the limbic system and other major structures and centers within the brain.

NEURAL CENTERS DURING MEDITATIVE STATES

In discussing possible neural mechanisms that might operate during states of meditation, Gellhorn comments on the findings of Das and Gastaut's early study of EMG, EEG, and pulse rate in subjects practicing a form of yoga meditation. Das and Gastaut, as we discussed in Chapter 3, reported the following EEG changes during meditation: first, an increase in the alpha frequency and a decrease in amplitude; then, the appearance of 15-30 cps activity; next, the appearance of even faster waves (40-45 cps) with amplitudes of 30-50 microvolts; and finally, a slowing of the alpha frequency to 7 cps. They also reported that during meditation the yogis showed an absence of EMG activity and an increase in pulse rate.

Gelhorn interprets muscular relaxation and lack of EMG activity under normal conditions to be a sign of parasympathetic activity and EEG synchronization. He suggests that a reduction in proprioceptive impulses causes a reduction of activity in the posterior hypothalamus and a release of anterior hypothalamic activity, which in turn leads to increased parasympathetic activity and EEG synchrony.

He comments that the state attained by the yogis is unique in that along with muscular relaxation and the resulting decrease in proprioceptive discharges, the EEG activity indicates an alert state of wakefulness. He points out, however, that there are conditions (such as in curarized animals) where, despite the loss of almost all skeletal muscle discharges, the animal will show EEG arousal if various cortical areas that act on the reticular activating system are stimulated. Gellhorn hypothesizes that "...reduction of the proprioceptive activation of the hypothalamus and the reticular formation in the relaxed state creates favorable conditions for cortical thought processes to go on without interference, while an adequate state of awareness is maintained through corticofugal impulses impinging on the reticular formation."

Anand and coworkers (1961), in explaining the increased alpha wave activity that occurred in several yoga meditators, also speculate about the neurophysiology of meditation. They postulate that the brain activity of yogis during meditation is dependent upon the mutual influences between the RAS and the cortex and is not dependent upon the activation of the RAS through external and internal afferents. In supporting their hypothesis they refer to the postulations of Garoutte and Aird (1958) that bilateral alpha and beta rhythms are probably under the control of a subcortical pacemaker or system of pacemakers.

MODELS OF THE NEUROPHYSIOLOGY OF TRANSCENDENTAL MEDITATION

In my Ph.D. thesis (Wallace, 1970a), I developed a preliminary model of the physiological changes that occur during the TM technique. In order to explain the increased slow alpha activity, which occurs specifically in the frontal regions of the brain, I referred to the work of Andersen and Andersson (1968), which suggested, contrary to earlier theories, that alpha activity was not controlled by a central pacemaker in the diffuse thalamic system but rather was produced and controlled by the various thalamic nuclei.

Each nucleus acts as a separate pacemaker, partially influenced by the diffuse thalamic system, but having the ability to produce rhythmic activity in the thalamus and in that area of the cortex to which it anatomically projects. The increase in alpha activity in the central and frontal areas of the brain might then be explained by a shifting of a "leading" pacemaker from a thalamic nucleus projecting to the occipital areas to a nucleus or nuclei projecting to the more frontal areas.

The increase in rhythmic activity and resulting increased integration of the frontal areas of the brain during TM is an appealing element in any model of higher states of consciousness. The frontal cortex of the brain is more prominent in man than in other species; it is also considered to be responsible for higher mental functions and for the temporal continuity and stability of the sense of self. The shift of focus of EEG rhythmic activity toward the frontal areas during the TM technique might be caused by an alteration in the activity of neural structures (such as the hypothalamus) that are anatomically connected to thalamic nuclei which project to the frontal and central regions of the cortex (such as the dorsomedial nuclei).

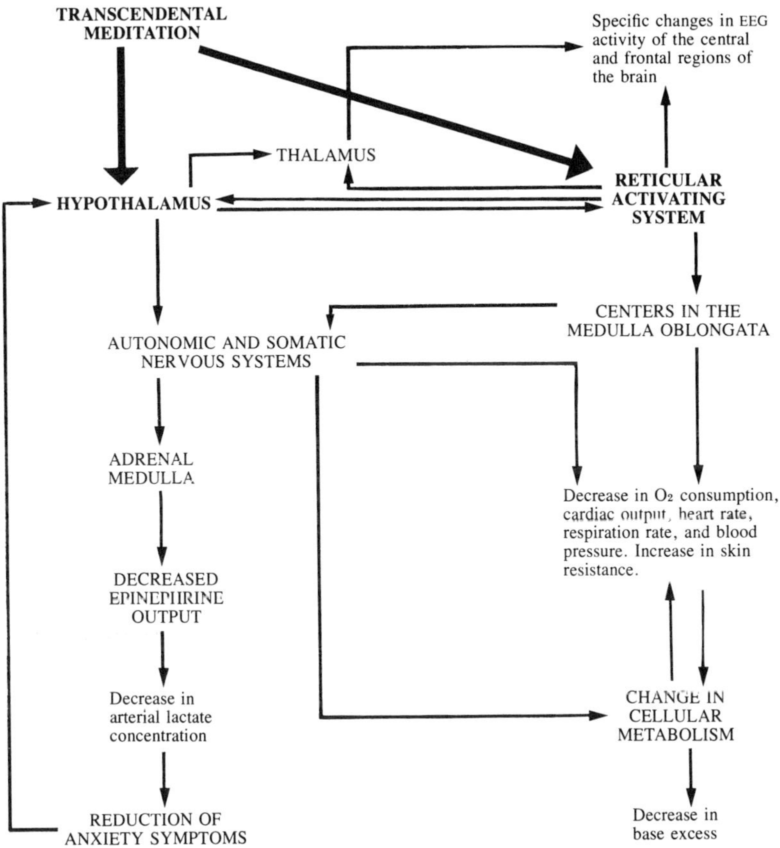

Wallace, R.K., *The Physiological Effects of Transcendental Meditation.* Ph.D. Thesis, Dept. of Physiology, University of California, Los Angeles, 1970.

In brief, I postulated that during the Transcendental Meditation technique ".... (a) The hypothalamus interacts with thalamic nuclei to facilitate specific alpha wave frequencies in certain areas of the cortex; (b) it interacts with the RAS to inhibit certain neural centers that act on the diffuse thalamic system and to decrease irrelevant sensory input on the thalamic nuclei; and (c) the hypothalamus either directly integrates autonomic and somatic activity or indirectly acts on the medullary centers through the RAS to produce or influence the changes seen in oxygen consumption, cardiac output, heart rate, blood pressure, skin resistance, and blood lactate concentration." I further suggested that there might be an alteration in cellular metabolism, which would then affect lactate production as well as oxygen transportation and utilization.

The most recent work by Jevning, Wilson and coworkers extended certain aspects of this model, particularly in the area of cellular metabolism. Their work in muscle metabolism clearly shows a marked decrease in oxygen utilization and an alteration in carbon dioxide production, and their work on lactate production in red blood cells suggests an alteration in metabolism due to the production of a new blood factor.

Banquet and Sailhan (1974), based on their extensive EEG research, suggest that since the EEG during TM resembles a kind of diffuse recruiting response, meditation must therefore activate the nonspecific thalamic nuclei while simultaneously blocking external input and synaptic inputs. The regular practice of the TM technique, they postulate, could be the basis of a better functional integration of specific and nonspecific areas of cortical and subcortical structures within the brain. They conclude that during TM there is "... a better integration at all levels of neural functioning—between perceptive and active functions, between the two hemispheres, and between cortical functions and subcortical visual-emotional functions."

Three Types of Electroencephalic Ordering

It has been shown by Dr. J-P. Banquet of France that the EEG signals (brain waves) induced during and after the practice of Transcendental Meditation are remarkably synchronous and coherent and indicate superior integration and coordination of different brain areas. (Banquet, 1972, 1973) Three distinct types of integration are implied by this observed brain wave synchronization.

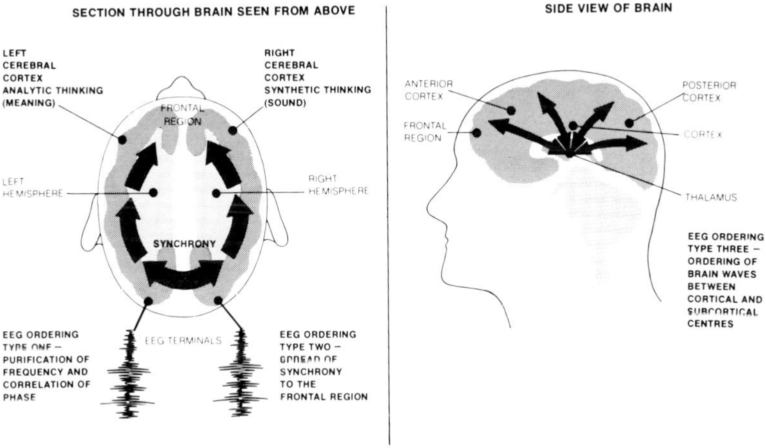

Banquet, Jean-Paul. 'EEG and Meditation'. *Electroencephalography and Clinical Neurophysiology* 33 (1972): 449.

Banquet, Jean-Paul. 'Spectral Analysis of the EEG in Meditation'. *Electroencephalography and Clinical Neurophysiology* 35 (1973): 143.

Another psychophysiological model has been proposed by Arenander (1980). This model considers the role of the mantra used during TM as a novel stimulus and attempts to describe the unique effects of the mantra in terms of certain brain structures which are critical to the flow of information processing.

In developing his model, Arenander refers to several neurophysiological structures in the brain which have been suggested, to be involved in information processing. The first of these structures is the hippocampus, which is thought to have an important role in short-term memory and stimulus recognition. Any incoming stimulus, prior to extensive processing, is apparently "matched" against stored stimulus representation within the hippocampus. If the stimulus is "matched" (i.e., is recognized), then another brain structure known as the basal forebrain is activated. However, if there is a mismatch of the stimulus in the hippocampus (i.e., it is not recognized), then a structure known as the mesencephalic reticular formation (MRF) is activated. The basal forebrain and MRF in turn influence the nucleus reticularis thalami (NRT). The NRT plays an important inhibitory role in the flow of information from the thalamus to the cortex. The MRF activates the NRT, while the basal forebrain inhibits its activity.

According to Arenander's model, during the TM technique the novel stimulus, the mantra, causes an activation of the MRF, thus opening the gates of the NRT to allow information to go to the cortex. When the mantra is no longer recognized as novel, the basal forebrain is activated, thus inhibiting the NRT and preventing information from continuing to the cortex for unnecessary processing. Activation of the basal forebrain causes the body to settle down. As the mantra is perceived in increasingly refined values, it is again mismatched and the cycle repeats. Thus are the dynamics of TM created, in which a settled and restful physiology coexists with mental alertness.

The settled or "de-excited" physiological state experienced during TM often persists to some degree after meditation, as evidenced by various physiological parameters as well as personal experience. The basal forebrain, according to Arenander's model, may still be exerting its inhibitory influence on the nervous system during the post-meditation period.

Some researchers have suggested that during the TM technique the resonant quality of the mantra may serve as a stimulus in the ideation speech area in the dominant temporal lobe. Since the TM technique does use the sound quality of thought as a means of allowing the attention to transcend to progressively finer levels of thought, it is appropriate to consider the involvement of the major brain areas of both hearing and language.

During the TM technique the experience of the more refined aspects of a specific sound or thought would necessarily involve some of the areas concerned with speech and verbal communication. Perhaps these areas, acting as a focal point of integration, initiate neurophysiological integration in other areas of the brain. This, in turn, might result in balancing and settling physiological processes throughout the body. The neurophysiological modeling of the process of transcending during TM is a rich area of investigation and continues to be pursued.

NEW DIRECTIONS IN RESEARCH ON PURE CONSCIOUSNESS—LEVEL 1 THEORETICAL MODELS

A number of new areas for future research have been suggested, in order to provide the data for developing more sophisticated models of higher states of consciousness. In considering these new areas of research we will focus on both their improvements in methodological design as well as their use of more detailed and advanced physiological measurements.

Previous studies have clearly shown the methodological advantage of identifying periods of pure consciousness during the TM technique in order to distinguish between physiological changes during TM and relaxation or rest. Farrow and Hebert's study, in which periods of breath suspension were highly correlated with subjective reports of pure consciousness, suggests that using periods of respiratory suspension as a marker of pure consciousness is effective at least in those subjects who show this type of pattern. Such an approach, of course, cannot be generalized to all subjects. However, this process of selecting advanced subjects who show this specific characteristic of breath suspension greatly facilitates the analysis of the data. In addition to more clearly defining the independent variable (i.e., isolating periods of pure consciousness during TM), it is also important that studies do not rely on dependent variables (i.e., type of measurement taken) that are not able to distinguish between different states of relaxation.

Simple measurements of autonomic functions such as heart rate or skin resistance alone do not appear to be either useful or valuable in identifying and studying the least excited state of consciousness. By far the most ideal type of design is one in which a multi-disciplinary approach is taken, in which a number of discriminatory physiological parameters are measured simultaneously. For example, simultaneous measurements of continuous respiration rate, oxygen consumption, EEG power and coherence (particularly in the frontal and central areas), along with various autonomic measures might be the minimal parameters necessary. Further, the analysis of data should take care not merely to average the data over the entire meditation period, but to analyze it according to the specific substate experienced as indicated from the various physiological parameters and subjective reports.

For example, an event marker could be used by subjects to indicate specific experiences of pure consciousness during the TM program. Also, standard EEG analysis along with subjective reports, if available, could be used to eliminate any periods of sleep that might naturally occur if the subject is overly tired or because of laboratory conditions. In this way, the experimenter would avoid making the mistake of averaging periods of sleep with periods of pure consciousness and thus not properly defining the independent variable in question—the least excited state of consciousness.

A finer analysis of the data, along with more sophisticated measures of localizing brain functioning, would also be important. This would be necessary in order to determine whether or not there were any internal rhythms or cycles during the TM technique corresponding to the subjectively reported experience of the alternation between periods of "transcending," in which thought activity is minimal, and "normalizing," in which there is a great deal of thought activity.

One obvious area of research is the expansion of EEG measurement to include a larger array of electrodes, along with more sophisticated types of analysis and displays such as topographical mapping and multivariable programs using EEG power, coherence and evoked potentials. While these approaches may indeed give a more detailed picture of the dynamics of cortical functioning during the TM technique, they too are limited in their ability to give a clear picture of localized neural activity indicated during behavioural states. It is possible that a far precise picture could be gained by the use of more advanced technologies such as positron emission tomography (PET scan) and nuclear magnetic resonance (NMR). Initial studies might be done on a clinical population that was already undergoing PET scans. Patients could be measured before beginning the TM technique, during its practice, and several months after regular practice.

Whichever of these procedures proves to be the most

useful, it will be important to carefully correlate measurements of brain blood flow and metabolism with other electrophysiological and biochemical measurements and markers during the TM technique. This will ensure that first, the state of pure consciousness (not other possible states) is being studied and second, that the fullest possible description of the physiological characteristics of the pure consciousness state is acquired.

Simultaneous biochemical measurements such as those undertaken by Jevning and Wilson in their studies of lactate metabolism would be valuable. Their approach of using a simple model system (i.e., lactate generation in the red blood cell) for the study of physiological changes during the TM technique is a particularly creative and useful example and should be followed by other researchers who wish to study detailed physiological processes. Again, if care could be taken to draw samples during periods that are associated with the state of pure consciousness (as determined by either physiological parameters or by subjective reports), the results should be far more meaningful and significant.

More biochemical studies in such areas as intermediate metabolism (i.e., pyruvate, glucose, etc.), endocrinology (pituitary hormones such as ACTH and TSH), and neurochemistry (levels of neuropeptides such as B-endorphin or neurotransmitters such as serotonin), would also be useful for elucidating the precise biochemical nature of the state of pure consciousness. In this regard, perhaps the most significant and important research being conducted now is the attempt to isolate substance M, the biochemical which could possibly be key to the experience of meditation and responsible for orchestrating the numerous physiological changes associated with the experience of pure consciousness.

The models we have discussed so far have emphasized the neurophysiological mechanisms during the TM technique. Of equal importance are the neurophysiological mechanisms responsible for the many effects seen in TM subjects outside of meditation. A number of TM researchers have proposed a developmental model of the growth of higher states of consciousness in which the repeated experience of pure consciousness results in a wide variety of beneficial mental and physical changes.

I suggested quite strongly in chapter 7 that this developmental process involves a specific reorganization of neurophysiological functioning which can be measured by a number of physiological parameters. In the same manner that experience plays a critical role in the developmental processes in animals, the experience of pure consciousness seems to produce both acute and long-term biochemical, metabolic and electrophysiological changes in human beings.

Kesterson (1986) has outlined the steps by which such a developmental process might occur. He suggests that the initial process involves the excitation of a specific set of inborn reflexes, specifically the orienting reflex, which in turn, as a result of the experience of pure consciousness, are generalized to form an integrated physiological response. The regular and repeated experience of pure consciousness, with its accompanying physiological changes, has a conditioning effect on the nervous system, resulting in specific trait effects which are seen outside of meditation. As the individual progresses to higher states of consciousness, these trait effects become more and more pronounced, such that completely different styles of neurophysiological functioning are produced. Each new state of consciousness is accompanied by a different pattern of physiological correlates, often quite distinct from the previous pattern. Thus, while there is a continual process of development, distinct phase transitions may occur as the individual experiences higher states of consciousness.

Already, our initial hypothesis concerning the pattern of physiological changes during and after the TM technique has been revised considerably, both with the addition of more advanced subjects and with more comprehensive measures. Based on the initial studies it was suggested that there was a decrease in sympathetic tone as a result of TM. The changes in blood flow distribution seen by Jevning and coworkers and the studies on norepinephrine levels by Lang and coworkers indicate, as we have noted before, that there may be an increase in both sympathetic and parasympathetic tone. The changes may involve more a balanced activation of different systems rather than a selective deactivation of one system.

Bujatti has suggested a similar model involving the balance of two key neurotransmitters, serotonin and norepinephrine. As a result of the regular practice of TM, Bujatti hypothesizes that both serotonin levels and norepinephrine levels increase in a balanced manner. The higher levels enable the system to make the transition from the extreme states of deep rest achieved during TM to states of alert activation outside of meditation.

Whatever the precise neurophysiological mechanisms responsible for the long-term changes resulting from the practice of the Maharishi Technology of the Unified Field, the overall changes do suggest a more coherent and balanced style of functioning which results in a number of improvements in health and behavior.

One of the most productive areas of TM research outside of meditation has been to study TM subjects during and after various challenge or stress tests. This type of design would be a good test of Bujatti's hypothesis, since it enables the researcher to see the degree and range of the physiology's adaptability. Previous studies have indeed shown a faster recovery from stressful stimuli. To extend the current research in this area, again a multidisciplinary approach would be most useful, using simultaneous measurements of several physiological parameters such as temperature, heart rate, blood pressure, blood flow, phasic skin resistance, and oxygen consumption, along with several biochemical measures like epinephrine, norepinephrine, serotonin, cortisol, and perhaps other hormones.

Either a cross-sectional design using non-meditators, short-term meditators, and long-term TM-Sidhi participants, or a prospective longitudinal design using subjects before and after starting the TM technique, and before and after starting the TM-Sidhi program, could be employed. In either case it might be initially desirable to establish a baseline period using continuous or frequent measurements over several days. In this way valuable data could be gained on the day-to-day stability of these measurements, as well as an accurate value of the subjects' basal or resting level before beginning the recovery test.

A further area of research along these lines would be long-term studies on the stability of circadian or ultradian rhythms. Such studies would necessarily involve fewer measures, but they would be particularly useful in extending current findings, which suggest that the TM and TM-Sidhi program acts as a synchronizer in establishing greater orderliness in biological rhythms. Long-term measurements of metabolic parameters would also be interesting in the light of previous research by Jevning and Wilson.

The assessment of physiological stability and adaptability as indices of the development of higher states of consciousness is especially relevant in the light of the beneficial effects of the TM and TM-Sidhi program on health and aging. For example, adaptability to a challenge or stress situation has been suggested to be important in the assessing of cardiovascular risk factors such as hypertension and type A coronary prone behavior. Greater longevity seems to be associated with greater homeostatic adaptability at an older age, along with more stable physiological rhythms.

The TM and TM-Sidhi program offers a unique model for the study of optimal health and longevity. A deeper understanding of the underlying physiological and neurophysiological mechanisms in the development of higher states of consciousness should help answer fundamental questions concerning physical and mental health and the aging process. Current research, particularly on TM-Sidhi participants, suggests that what we have previously considered to be the normal value for certain physiological parameters at a given age may in fact be abnormal, and that with the development of higher states of consciousness, entirely different normal and optimal values may be established.

An important part of future studies should be a very careful analysis of subjective experiences in order to delineate the physiological and psychological characteristics of each specific state of consciousness described by Maharishi. The preliminary sleep studies undertaken by Banquet on advanced subjects experiencing characteristics of the state of cosmic consciousness (i.e., witnessing during sleep) could be extended. One final, more ambitious suggestion for the longitudinal evaluation of higher states of consciousness would be use multivariable measurements in a prospective study comparing individuals starting the TM and TM-Sidhi program with controls over a longitudinal period of 20 to 40 years, preferably beginning the study early in the individuals' lives.

THEORETICAL MODELS—LEVEL 2

Since Maharishi's introduction of the TM technique to the West, he has encouraged scientists to develop more sophisticated models of consciousness incorporating the latest findings of modern physics. One of the pioneers in this field is Dr. Lawrence Domash. In an extremely stimulating article, Domash (1975) developed a model which offers an entirely new approach to understanding the neurophysiology of higher states of consciousness in terms of quantum physics.

As a rationale for developing this model, Domash points out that consciousness can be viewed as a variable rather than as an on-off property, which is capable of being decreased in drowsiness and sleep, and increased or "expanded" in at least one distinct stage of the waking state, i.e., pure consciousness. This view suggests that the nervous system is capable of a sequence of phase transitions in which consciousness can be considered in terms of a long range correlation effect in the brain measured through a physical order parameter.

This model of consciousness is based upon considerations derived from the most basic properties of physical systems. In a physical system, the degree of purity or orderliness depends upon the temperature or activity of the constituents of the system. For example, in its most excited state, water exists in a gaseous form as steam; as we reduce the temperature, the water undergoes a phase transition, first to a liquid state and finally to a solid state in the form of ice. In its solid state it is extremely orderly and pure.

A more profound example of the principle of reducing activity and thus increasing orderliness and purity (known as the Third Law of Thermodynamics) is seen in the phenomena of superfluidity and superconductivity. If the temperature of liquid helium is decreased to a few degrees above absolute zero (-273 K), an extraordinary phase transition occurs, producing a unique fourth state of matter—different from the gaseous, liquid, or solid states—known as the superfluid state.

In this superfluid state, the helium can escape through

glass containers; it becomes as if unbounded. Its resistance to flow is zero, and its capacity to conduct heat is infinite. These unusual properties result, we learn from physics, from the high degree of orderliness or coherence that develops among the helium atoms. They behave as a single helium atom; the quantum mechanical properties of the atom are allowed to manifest on a macroscopic scale.

If the temperature of a metal, such as lead, is reduced to a few degrees above absolute zero, a similar type of phase transition occurs, resulting in the phenomenon of superconductivity. The electrons within the material form themselves into pairs, which become coherent with one another in the sense that their underlying wave nature, as described by quantum mechanics, becomes perfectly synchronized. As in the case of helium, a coherent structure is formed such that all the parts behave collectively as an integrated whole. The underlying orderliness of the microstructure of matter, which is normally hidden, now manifests again on a macroscopic scale. One surprising and previously unpredicted property of superconductors is that an electron current, when started in a superconductor metal loop, will flow forever.

The superfluid and superconductive states thus illustrate how it is possible, at least in a physical system, to attain the ground or least excited state of matter even on a macroscopic scale. These states also reveal to us the existence of extraordinary properties within matter, which would never have been realized without the technology to produce the superfluid or superconductive state.

Domash has employed these models of superfluidity and superconductivity to describe the experience of pure consciousness. He suggests that the degree of consciousness may be related to the degree of long-range spatial and temporal ordering of the vast collection of neurons within the brain. We may think of the TM technique as a means for systematic de-excitation of the nervous system, lowering its "mental temperature," while preserving conscious awareness. At the level of least excitation of consciousness, a type of phase transition occurs to a distinct and more highly ordered state, stabilized temporally and highly correlated spatially, and giving rise to the subjective experience of "pure" consciousness.

When considering a living system it is inaccurate, of course, to consider phase transitions of the strictly Third-Law type; rather it is more appropriate to consider non-equilibrium phase transitions. However, as Domash points out, there are overriding similarities and close mathematical analogies between the style of phase transition which actually results from non-equilibrium dynamics, such as that which occurs in laser light, and the type which results from low temperature (Third Law) effects, such as the transition from paramagnetism to ferromagnetism. He refers to the work of W. Little, who has developed a mathematical model, analogous to the Ising Model in ferromagnetism, of a network of neurons that does exhibit phase transitions to ordered states in the form of persistent firing patterns (Little, 1974).

Domash suggests that the type of phase transition which occurs during TM is a direct result of a kind of macroscopic quantum behavior within the nervous system. He refers to the work of a number of previous researchers suggesting the appearance of macroscopic quantum mechanical behavior in biological systems. On theoretical grounds, a number of physicists and neuroscientists have already suggested that the function of the brain may be inherently quantum mechanical. For example, Bohr (1961) suggested that thought involved such small energies in the brain that it may be governed by quantum effects. Stuart and coworkers (1979) have further attempted to model memory and retrieval in the brain in terms of quantum mechanical systems, and E.H. Walker (1970) has suggested quantum mechanical tunneling at the synaptic gap. Wigner (1962), in his work on the quantum theory of measurement, points to an intimate and unavoidable relationship between the quantum mechanical wave function and human consciousness.

One of the first scientists to propose that macroscopic quantum mechanisms such as superfluidity might be important for living systems was Fritz London (1961). Although no firm evidence for the existence of room temperature superfluidity has been discovered as yet, there has been a great deal of interest in this area (Cope, 1974). Domash points out that the most straightforward proposal for quantum macroscopic behavior in biological organisms would be electron superconductivity; however, it must of course involve new mechanisms different from the lattice phonon attractive interaction between electrons, which is responsible for ordinary low temperature superconductivity.

Domash suggests, as does E.H. Walker, that the site most appropriate to look for quantum effects might be the synapse. Walker, however, explicitly excludes the possibility of a macroscopic superconductivity-type wave function. Domash, on the other hand, suggests that something like the A.C. Josephson effect might take place across the synapse during the TM program, such that superconductivity in one neuron would be phase coherent with another by virtue of quantum tunneling. Domash further points out that the AC Josephson effect for a 70 millivolt synaptic gap potential would require the presence of electromagnetic oscillations in the microwave range.

The existence of a superconductive AC Josephson effect in living cells has also been suggested by Beischer (1971), who argues that the known magnetic field sensitivity of certain birds is difficult to explain by any physical mechanisms other than a biological Josephson effect magnetometer. As Domash suggests in his model,

> If localized regions of superconductivity are a common mechanism in cells, including neurons, then we suggest that the essential effect of the TM technique is to bring about phase correlation of these localized regions resulting in expansion of the wave function from molecular to truly macroscopic (approximately 1 cm) dimensions, accounting perhaps for the subjective experience of the sudden "unbounded infinite expansion" of the self awareness during the Transcendental Meditation technique.

Domash extends this model to higher states of consciousness such as cosmic consciousness. In terms of a macroscopic, quantum phase coherent model of pure consciousness, the state of cosmic consciousness is very reminiscent of the two-fluid models used to describe a superfluid. In this model there are two distinct interpenetrating fluids. One is in a macroscopic quantum state of perfect orderliness, while the other is in a partially excited state, carrying whatever disorder exists in the system. As Domash says.

> That this two-fluid description discovered independently by modern physics coincides so closely with the classic picture of the psychology of "enlightenment"... held for thousands of years to be the most ordered state of the mind and highest achievement of human life, is an especially elegant and satisfying point. If the connection bears scientific fruit, full credit is due to the TM technique and its principal exponent, Maharishi Mahesh Yogi, for having made this class of experiences accessible to scientific investigation for the first time.

These theoretical models of pure consciousness and higher states of consciousness provide a rich framework for a deeper and more profound understanding of the dynamical processes that underlie modern biology and modern physics.

More recently, Dr. John Hagelin has developed a model of consciousness incorporating the most current unified theories of physics. We have already considered this model in detail in Chapter 2. The integration of our current understanding of the unified field theories from modern physics with the ancient knowledge of Vedic Science as revived by Maharishi provides a remarkable new approach to uncovering the intimate relationship between matter and consciousness, mind and body. Further, it provides the foundation for a new and far richer comprehension of the neurophysiological processes supporting the growth of higher states of consciousness.

FUTURE RESEARCH—LEVEL 2
THEORETICAL MODELS

The most important and practical area of future research on the Maharishi Technology of the Unified Field is undoubtedly the study of the field effects of consciousness. The studies undertaken so far verify empirically the significant positive effect on society of having either one percent of the individuals in any community practicing the TM technique (Maharishi Effect) or the square root of one percent practicing the TM and TM-Sidhi program in a group (Super Radiance Effect). The theoretical basis of the Maharishi and Super Radiance effects must be considered from the point of view of quantum field theory. Further research in this area could help to precisely define the specific characteristics of the field effect produced. More important would be the development of a complete understanding of the neurophysiological basis of the Maharishi and Super Radiance Effects.

An excellent direction for future research would be to extend the earlier studies using physiological parameters which might be sensitive to the field effects of consciousness. Further, experiments should be designed showing how the field effects of consciousness vary with distance.

Most important, perhaps, would be to extend the previous findings on the Maharishi and Super Radiance Effects on a larger scale. An ideal type of study would involve the participation of an entire country. Not only could numerous scientific studies be done to measure the effects of the Maharishi and Super Radiance Effects on all areas of that society, but more importantly, very practical results could be observed in terms of the coherence and progress of the country as a whole. While this proposal may sound ambitious, a technology as powerful and extensively studied as the Maharishi Technology of the Unified Field merits a large scale demonstration program, especially in view of the critical situation in the world today. No other area of research in any field of science offers mankind such promising benefits.

Chapter 10

EDUCATION FOR ENLIGHTENMENT

> With all the glory of the present system of education and the rapidly increasing technology it has allowed man to develop; with all the ability of modern man to look into the great picture of the universe and know something of the vast galaxies; and to penetrate to the other extreme of life and know the mechanics of the world of the atoms and electrons, and use them for good; with all this vast knowledge, what is lacking? Completeness of knowledge. The thirst for knowledge is not satisfied. A man knows something of history, something of economics, something of physics or chemistry. But the field of knowledge is so vast; it is not possible for every man to have all knowledge. However, the total value of knowledge can be grasped on the level of one's awareness if the conscious mind is sufficiently expanded.
>
> — Maharishi

Partial knowledge produces fragmented and limited results, which are only partially successful. However great the benefits of the scientific advances of this age, we still possess only fragmented knowledge of the laws of nature. The result of this incomplete knowledge can be seen in the unwanted side effects that have been created and perpetuated by technological progress, culminating in the fear and constant threat of nuclear annihilation.

The age of modern science and technology has caused a dangerous imbalance in nature, and it must be corrected. We must progress from a fragmented knowledge of the laws of nature to a unified knowledge of the laws of nature. The problems and weaknesses found in virtually every society, whether we consider technologically advanced or developing countries, can be analyzed in term of the failure of the educational systems to culture the full potential of the human mind, to give complete knowledge.

"The purpose of education," according to Maharishi, "is to culture the mind of man, so that he can accomplish all his aims in life. For education to justify itself, it should enable a man to use the full potential of his body, mind and spirit. It should also develop in him the ability to make the best use of his personality, surroundings, and circumstances so that he may accomplish the maximum for himself and others."

In order to accomplish this purpose of education it is necessary to incorporate within the many systems of education today a technology of consciousness, through which each individual can directly experience the unified field of all the laws of nature, pure knowledge in its most complete state. Since the expansion of consciousness involves a progressive refinement of neurophysiological functioning, complete education involves a systematic development of the brain and the entire physiology.

We don't normally think of education as a physiological experience, but it clearly is. As a result of our educational experience a pattern or style of physiological functioning is acquired. Unfortunately, today the style most fostered by modern education is a highly imbalanced one in which the individual learns to continually overtax and stress himself. The result is clearly seen in the high prevalence of mental and physical stress-related diseases, which we have already discussed. We are literally conditioning our children in a style of physiological functioning that forces them to consume large quantities of medication—tranquilizers and other drugs—in order just to cope with the stress of life. The individual is under extraordinary pressure, and this is reflected in the breakdown of life-supporting values within our families, our communities, and our society.

There is a critical need to make education more complete—to establish as the foundation of education a program by which the physiology is properly cultured such that each individual is able to unfold his full creative potential, live in perfect health, and spontaneously act in accordance with natural law. The Maharishi Technology of the Unified Field offers such a program to education. In this final chapter we will examine the development of this integrated system of education. First we will examine it from a historical perspective, and then we will consider its practical application throughout the world.

THE INNATE PHYSIOLOGICAL ABILITY TO EXPERIENCE PURE CONSCIOUSNESS

The universal application of the Maharishi Technology of the Unified Field in education results directly from its use of an innate physiological ability present in every individual, regardless of his educational or cultural background. We all have the ability to produce the integrated physiological response seen during the TM and TM-Sidhi program because it is part of the genetic endowment of our nervous system. While we do not yet know the precise details of how this mechanism works, we do know that there exist neural pathways and structures that are capable of being activated during the TM and TM-Sidhi program and that support the experience of the least excited state of consciousness, the simplest form of human awareness.

We can surmise that if such neural pathways exist as a

part of the genetic endowment of all individuals, there should be historical records of the use of this innate physiological capacity to support the state of pure consciousness. Indeed, there are such records, particularly in the East in the ancient cultures of India, China, Tibet, and Japan. Very refined methodologies involving physical and breathing exercises as well as various types of mental techniques of meditation were developed in order to systematically culture and continually activate and reinforce the neural pathways that support the pure consciousness state.

Descriptions of the experiences involved in these processes of physiological refinement are primarily in terms of characteristics of the mental state ("unbounded bliss," the "immortal Self," the "ocean of consciousness moving within itself," etc.). However, there is often a corresponding description of physiological characteristics (suspension of breathing, the body being completely silent, the attention or awareness being more orderly and inwardly alert).

As part of their educational system, the ancient Eastern cultures incorporated these methodologies to refine physiological functioning. Some aspects were taught to almost all children at an early age, but the most extensive and profound teachings were given to the monks, who were revered and highly honored as the wisest and most profoundly educated in the culture. They were engaged in a lifelong continuing educational program.

Unfortunately, in most of these cultures what has survived of these practices and is taught today in the name of meditation involves rigorous concentration and contemplation and is not very effective in producing the integrated physiological response that helps support the pure consciousness state. As a consequence, few individuals actually gain the experience of higher states of consciousness, and the practices are regarded as impractical and esoteric.

The ability to experience pure consciousness is an innate capacity, like the ability to speak a language. However, in learning a language, the individual must have the actual experience of speaking it; in learning to transcend the more excited levels of thinking, the individual must be exposed to the actual experience of pure consciousness, and he must be taught in some systematic manner. What has been missing from our educational systems is the opportunity to systematically learn to experience this most fundamental physiological state.

We live in a world that has come to accept a limited view of the infinite resource of human intelligence. The genius is regarded as the exception; man is considered a victim of his environmental conditions, and suffering is viewed as a natural part of life. We have, in spite of all our enthusiasm to explore nature's wonders, circumvented the very greatest of these—our own inner Self. Yet if we look to the personal experiences of some of the greatest leaders, poets and scientists in history, we find very vivid and real accounts of this state.

In the writings of the early Greeks, at the dawn of Western civilization, there are many accounts in which the nature of life was understood in terms of an underlying source of Being. Perhaps the most extensive examination of how to live life in harmony with the laws of nature is given in the dialogues of Socrates. In Plato's Republic, the very basis of Western thought, Socrates clearly describes the source of all knowledge, "the Good," as being beyond sensation, beyond intellectual thought. In order to live life in accord with natural law one must just experience "the Good." Unfortunately, the actual technique Plato gave to experience "the Good" involved quite rigorous training. Thus, while Plato has been extensively studied, his demand that "the Good" must be experienced as the basis of any education has been misunderstood and never practically applied.

Throughout the history of Western thought the Platonic theme of "know thyself" has recurred in the writing of great individuals. It is clear that while no systematic procedure was available for transcending intellectual thought to the state of pure consciousness, nevertheless occasionally some few have at least glimpsed this experience.

Marcus Aurelius writes in *Meditations:*

> Man seeks seclusion in the wilderness, by the seashore or in the mountains—a dream you have cherished only too fondly yourself. But such fancies are wholly unworthy of a philosopher since at any moment you chose you can retire within yourself. Nowhere can man find a quieter or more untroubled retreat than in his own soul. Above all he who possesses resources in himself, which he need only contemplate to secure immediate ease of mind. Avail yourself often then, of this retirement, and continually renew yourself.

Even more elegant are the accounts given by Alfred Lord Tennyson and William Wordsworth. Tennyson writes:

> ...all at once, as it were out of the intensity of the consciousness of individuality, individuality itself seemed to dissolve and fade away into boundless being, and this not a confused state but the clearest of the clear, the surest of the sure, utterly beyond words—where death was an almost laughable impossibility—the loss of personality

> (if so it were) seemed no extinction, but the only true self. I am ashamed of my feeble description. Have I not said the state is utterly beyond words.

Wordsworth, in his poem "Tintern Abbey," gives a vivid description of the experience of transcending, even using physiological terms to characterize the state:

> ...that serene and blessed mood,
> in which the affections gently lead us on—
> Until, the breath of this corporeal frame
> And even the motion of our human blood
> Almost suspended, we are laid asleep
> In body, and become a living soul;
> While with an eye made quiet by the power
> Of harmony, and the deep power of joy,
> We see into the life of things.

One of the richest descriptions is given by Albert Einstein:

> There are moments when one feels free from one's own identification with human limitations and inadequacies. At such moments one imagines that one stands on some spot of a small planet gazing in amazement at the cold yet profoundly moving beauty of the eternal, the unfathomable. Life and death flow into one and there is neither evolution nor destiny, only Being.

These experiences are but a few of those recorded in our Western tradition. They indicate to us that it is possible to spontaneously transcend the more excited levels of thinking and to activate the neural pathways that will support the subjective experience of pure consciousness. Unfortunately, this experience may be only momentary and, without the proper technology to systematically transcend to lesser excited states of awareness and refine neurophysiological functioning, it may never occur again.

Wordsworth was forever trying to regain this experience. After years of trying to force it to recur, he became frustrated and began to doubt its validity. His experience unmistakably demonstrates the need for both experience and knowledge. It is not enough to once allow the awareness and physiology to slip into the pure consciousness state. The experience must be accompanied by knowledge both about its nature and about how to systematically regain it. This knowledge and experience of pure consciousness, as well as higher states of consciousness, has been missing from education. It is Maharishi's great genius not only to revive the knowledge and experience of consciousness in its complete form, but also to make it accessible to millions of active people throughout the world.

REVIVAL OF KNOWLEDGE

Education in traditional Vedic culture, as described by Maharishi, involved a very intimate teacher-student relationship. Just as a graduate student in science who wishes to become a skilled researcher must train under a senior professor, the Vedic student's "trained" under the expertise and guidance of an enlightened "master." The Vedic student's inward exploration into the nature of consciousness employed a specific set of laboratory procedures and a very specialized tool of investigation. The laboratory was the mind, the procedures were the techniques of meditation, and the specialized tool was the nervous system in its most integrated and refined state.

In order to experience the more refined levels of consciousness, the student had to learn how to turn his senses inward in such a manner that the awareness was allowed to transcend to quieter, more subtle levels of the thinking process, until it transcended thought altogether and reached the state of pure consciousness, without any thoughts or objects of perception. As with the scientist, experience was the critical factor. The methodology used was based upon experimentation, not belief or moodmaking. The experience had to be reliable, repeatable and accessible to others. The Vedic hymns of India serve a function similar to the modern journals of physics or biology. They are records of experiments, to be used by all those who want to replicate and re-verify the experiment.

Why, we might ask, haven't scholars and scientists of the past made use of these records; why are they only now being rediscovered? Why was the knowledge of this highly sophisticated research program into consciousness lost over time?

First, because the human nervous system is an exceedingly delicate instrument with many possible physiological states, verbal or written instructions were not always sufficient to preserve the more subtle aspects of the knowledge. It is necessary to have at hand a living teacher able to correctly interpret the instructions on the basis of his own experience.

According to Maharishi the knowledge of higher states of consciousness has been completely revived and then lost, not once, but several times throughout the history of man. In the following excerpt from the preface to his commentary on the Bhagavad-Gita, Maharishi cites a classic example of the revival of this knowledge by Shankara, one of India's greatest teachers, and then its inevitable loss over time.

> Shankara not only revived the wisdom of integrated life and made it popular in his day, but also established four principal seats of learning in four corners of India to keep his teaching pure and to ensure that it

would be propagated in its entirety generation after generation. For many centuries his teaching remained alive in his followers.... But in spite of all his foresight and endeavors, Shankara's message inevitably suffered with time the same misfortunes as those of the other great teachers.

....The state of Reality, as described by the enlightened, cannot become a path for the seeker, any more than the description of a destination can replace the road that leads to it. When the truth that Being forms the basis of the state of enlightenment became obscured, Shankara's statements about the nature of the goal were mistaken for the path to realization.... This is the tragedy of knowledge, the tragic state that knowledge must meet at the hands of ignorance. It is inevitable, because the teaching comes from one level of consciousness and is received at quite a different level. The knowledge of Unity must in time shatter on the hard rocks of ignorance. History has proved this again and again. Shankara's teaching could not prove an exception to the rule.

....It was the perfection of his presentation that caused Shankara's teaching to be accepted as the core of Vedic wisdom and placed it at the center of Indian culture. It became so inseparable from the Indian way of life that when, in course of time, this teaching lost its universal character and came to be interpreted as for the recluse order alone, the whole basis of Indian culture also began to be considered in terms of the recluse way of life, founded on renunciation and detachment.

> When this detached view of life became accepted as the basis of Vedic wisdom, the wholeness of life and fulfillment was lost. This error of understanding has dominated Indian culture for centuries and has turned the principle of life upside down. *Life* on the basis of detachment! This is a complete distortion of Indian philosophy. It has not only destroyed the path of realization, but has led the seekers of Truth continuously astray.
>
> ....Interpretations of the Bhagavad-Gita and other Indian scriptures are now so full of the idea of renunciation that they are regarded with distrust by practical men in every part of the world.

The present revival of the complete knowledge and experience of Vedic Science is, according to Maharishi, a result of the need of the time. When natural law declines to the point that it is constantly being violated by everyone in the society, and the entire social fabric begins to disintegrate, then nature responds by reviving the knowledge and experience of higher states of consciousness.

According to Maharishi, in our era nature's response came in the form of the living example of the wisdom and fullness of enlightenment. This embodiment of wisdom was Maharishi's teacher, His Divinity Swami Brahmananda Saraswati Jagad Guru Bhagwan Shankaracharya of Jyotir Math, referred to by Maharishi as "Guru Dev." Guru Dev was from 1941 to 1953 the Shankaracharya of Jyotir Math in the Himalayas (a seat of Vedic leadership that traces its origin directly from Shankara and in turn from the long tradition of masters dating back to the

Vedic culture of ancient India). The current revival of the knowledge of Vedic science dates from when Guru Dev accepted this seat. Maharishi gives full credit to his master, Guru Dev, for this revival.

Maharishi's education under Guru Dev was traditional and complete. Dr. Lawrence Domash, in his introduction to *Scientific Research on the Transcendental Meditation and TM-Sidhi Program: Collected Papers, Vol. I* (1977), gives a beautiful description of this education.

> Maharishi's background in India is a most pure and classical one. In the West, research scientists have a habit of referring to a young Ph.D. in terms of his "master"—"so-and-so, who was a graduate student of so-and-so." The same custom has held true from time immemorial in the tradition of Indian philosophy, since success in terms of that tradition has always been measured not by a man's intellectual attainment alone, but by the quality of his own inner life. Maharishi was the closest student of a great and famous teacher, a contemporary saint. He had the fortune, as an educated young man, to be accepted as a student of Swami Brahmananda Saraswati, who, although little known in the West, was widely recognized as the greatest of the purely spiritual preceptors of modern north India and was revered for many years by the community of Indian philosophers as a rare and perfect example of that peak of inner development whose ideals are extolled in the Vedas and Upanishads of ancient India.
>
> The education Maharishi received in this environment contained no element of

> the twentieth century; it could just as well have taken place two or ten thousand years before. But the uniqueness of Maharishi's situation was a truly modern one. On the one hand, Maharishi stood as the closest, most devoted, and most successful disciple of a teacher who had developed his own quality of mind to a degree of perfection so rare as to be almost extinct in today's world, including India. From him, Maharishi absorbed this timeless value, a breadth and delicacy of awareness that allows the full range of states of consciousness available to the human nervous system to be directly experienced in a systematic way. On the other hand, Maharishi brought to his studies a thoroughly contemporary spirit of inquiry, experimentation, logic, verification, and creativity—in other words, a scientific attitude. Maharishi's unique combination of qualities and circumstances has been most fortunate for the world of science, for through it he has been able to provide the missing bridge between the oldest tradition of human knowledge and the newest.

It is indeed "Maharishi's unique combination of qualities and circumstances" that has been most fortunate for Western science. Maharishi's great contribution has been to first revive the knowledge in its completeness and second to present it in a scientific and acceptable manner. In the late 1960's and early 1970's, as the result of the demand of many leading educators and scientists all over the world, Maharishi formulated the first truly holistic educational program, the Science of Creative Intelligence. More recently Maharishi has enriched this program with the development of the Technology of the Unified Field as the basis for an integrated system of education.

THE SCIENCE OF CREATIVE INTELLIGENCE

The Science of Creative Intelligence (SCI) is the study of the intelligence of nature and its accessibility to man in the simplest state of his own awareness. SCI explains that intelligence has two aspects, expressed and unexpressed, manifest and unmanifest. The expressed level is seen clearly in the orderly behavior of the laws of nature and in man's ability to understand these laws. The unexpressed level is seen as the pure potentiality of creative intelligence or pure intelligence, the field of pure consciousness. SCI provides the knowledge of these two levels both in terms of modern science and Vedic Science.

From the point of view of Vedic Science, SCI defines consciousness in terms of the self-referral nature of intelligence, through which the field of pure intelligence becomes aware of its own existence. Thus, the ability to know itself and become consciousness is inherent in the nature of pure intelligence. This self-referral quality is the basis for the dynamism of the state of pure consciousness and for all the mechanics of creativity in nature. Maharishi succinctly summarizes these mechanics in one of his lectures on the Science of Creative Intelligence: "When existence becomes conscious, intelligence becomes intelligent, and it is this intelligent aspect of intelligence that prepares it to play the role of creative intelligence."

The dynamic flow within pure consciousness is, according to SCI, the source of evolutionary processes. When consciousness begins to be conscious of itself, impulses or virtual fluctuations are created. We say "virtual" because the fluctuations take place within the unmanifest nature of consciousness, not in the field of manifest differences. These internal fluctuations or impulses of intelligence are described as the most fundamental impulses of one laws of nature, the very fabric of the unmanifest; they are the origin of space, time and direction. Pure consciousness reverberating within itself then begins to manifest into particular tendencies, and these tendencies give rise to the evolution of creation.

As we have shown in chapter 2, this description of the mechanics of natural law unfolding from the field of pure intelligence is remarkably similar to the description of the unfoldment of natural law in unified field theories of modern physics. In the process known as sequential dynamical symmetry breaking, the pure unmanifest nature of the super-field, as a result of its inherent property of self-referral, spontaneously generates the four basic forces of nature.

In the field of developmental physiology the unfoldment of the very compact and unexpressed knowledge within DNA occurs through a similar process of self-referral, in which DNA interacts with itself in the sequential expression of genetic information. These and other examples clearly illustrate the parallel description of the mechanics of creation as given by modern science and Vedic Science. SCI combines these two very different approaches to knowledge into an integrated science.

The understanding of the mechanics of creativity given by SCI, however, is not on a purely abstract, theoretical basis but in fact includes a technology, the Maharishi Technology of the Unified Field, which enables the individual to directly experience the laws of nature in their unmanifest, "seed" form. SCI thus includes two very different methodologies to experience natural law, the objective approach of modern science and the subjective approach of ancient Vedic Science. At a symposium in 1971 at the University of Massachusetts at Amherst, Maharishi clearly distinguished between these two methodologies of gaining knowledge:

> The Western scientific approach to knowledge is based on the nonvariability of the objective means of observation. The

> Eastern approach to knowledge is based on the nonvariability of the subjective means of observation. Speaking for the East, we know there is a level of consciousness called "ritam bhara pragyan," which is nonvariable in its nature; and therefore, on that level, the knowledge of an object never changes, remaining authentic and truly scientific for all time. Western science has continued to contribute to the advancement of civilization because good minds in the West have persistently applied the objective methodology of gaining knowledge, whereas Eastern civilization has not continued to contribute to the advancement of civilization because good minds in the East have not persistently applied the subjective methodology of gaining knowledge.
>
> Rig Veda proclaims that knowledge is structured in consciousness. Now the teaching of Transcendental Meditation—in the structure of the Science of Creative Intelligence—enriches the knowledge of the West and revives the knowledge of the East and will give the benefit of both methodologies to every man in the world. The Science of Creative Intelligence is the glorious meeting ground for the Eastern and Western ideals of gaining knowledge. All men everywhere will enjoy the highest ideal of life in fullness of all glories, material and spiritual.

These two separate methodologies of gaining knowledge are by no means exclusive. As Maharishi points out, through SCI they can be integrated into a far more powerful methodology of gaining knowledge, which incorporates the advantages of both.

Einstein, along with other great scientists, was noted for his intuitive abilities, his deep personal inner feeling for "the way nature works." The creative genius of these great minds was founded on this ability to combine intellectual analysis with the intuitive faculty. Yet, while many writers have noted and commented on this phenomenon and its importance for scientific investigation and for learning in general, our educational systems have until now been unable to provide such an integrated approach for all students.

Learning consists primarily of assimilating knowledge and perfecting logical analysis. There has been no real methodology for transcending to more unbounded states of consciousness and spontaneously culturing the intuitive ability to understand and live in accord with natural law. Maharishi's introduction of the Science of Creative Intelligence in the early 1970's marked a great step in the history of education, for the first time providing the foundation for a truly complete system of education.

MAHARISHI INTERNATIONAL UNIVERSITY

One of the natural consequences of introducing SCI in various educational systems in the U.S. and throughout the world was the desire to create an ideal educational setting in which SCI would be the foundation of the entire educational process. Maharishi International University was founded in the summer of 1971, following several large symposia on the Science of Creative Intelligence during which Maharishi spoke with eminent scholars, Nobel laureates, and leading educators. Among the thousands who attended the symposia were a number of professors from a wide variety of academic disciplines.

On the basis of their own experiences through the TM program and their insights into how the theoretical principles of SCI were found in their own disciplines, it occurred simultaneously to many that Maharishi's thinking, together with the technology of the TM technique, could be the basis for an ideal system of higher education. These professors became the founding faculty of MIU. Working closely with Maharishi, they spent the next few years developing a system of education that incorporated the very best of traditional higher education with the profound knowledge and experience provided by the Science of Creative Intelligence.

It had already been widely recognized that the current system of higher education had certain inadequacies and shortcomings. First, universities were the repositories of fragmented knowledge of many unrelated specialties, which seemed essentially irrelevant to most students' lives. Second, the universities had proven their inability to develop students who actually lived lives free from suffering and error; the type of education that was being offered was simply insufficient to prepare one for a life of success and happiness. Finally, demands for interdisciplinary education had not been successfully met; all the attempts to do so had been on a superficial level and had only diluted the boundaries between disciplines and weakened their integrity. Clearly, something new was needed, and the Science of Creative Intelligence satisfied this need in all respects.

In developing the MIU curriculum, SCI was introduced as a first science, a science and technology of consciousness that provided systematic analysis and direct experience of consciousness in its abstract, unmanifest aspect. SCI brought fulfillment to each discipline by showing that its intrinsic direction of growth already points in the direction of enlightenment. SCI identifies the natural desire for completeness in each discipline and shows that it can actually be fulfilled by bringing that area of study in contact with the field of pure consciousness. SCI thus completes each of the academic disciplines, and makes them capable of fulfilling the entire range of human evolution through a particular channel of knowing. This is particularly true in fields such as psychology, education, and sociology, where the subject matter has always been a search for wholeness in individual and collective life; this quest can only now be satisfied in a realistic way through evolution to higher states of consciousness. No academic discipline is actually complete until the Science of Creative Intelligence is an integral part of it.

Finally, SCI has provided a natural means to organize an academic curriculum along interdisciplinary lines without in any way spoiling the integrity of individual disciplines. It does this by relating the disciplines through their common value, the value of consciousness, the only fundamental feature that all branches of knowledge can hold in common. At MIU all the traditional fields are taught in their own terminology, without subtracting anything; and yet a genuine and satisfyingly holistic focus is created simply by adding the study of intelligence to each field.

SCI would not be successful if it were merely an intellectual analysis of the abstract principles of the growth of knowledge. The real power of SCI lies in students' experience of the unboundedness of intelligence within themselves through the TM and TM-Sidhi program. Through this experience they naturally become open to and appreciative of all academic disciplines; they become clear-minded and eager to learn because the relevance of knowledge to their own life is never out of their awareness. The ultimate value of SCI as a tool in the curriculum is best expressed by the motto of MIU: "Knowledge is Structured in Consciousness." By opening the awareness of the students to pure knowledge, pure consciousness, every field of investigation becomes as intimate to them as their own lives, and there is nowhere in the universe of knowledge where they do not feel at home.

The educational program at MIU stands as an ideal

model for all other universities. MIU has grown rapidly since its first years. Its main campus was established in Fairfield, Iowa, in 1974; a branch campus opened in Washington, D.C., in 1982. MII was accredited in 1980, and by 1984 had developed a number of master's programs, as well as doctoral programs in neuroscience, physics, physiology and psychology. MIU has been evaluated by many leading educators, who always remark upon the high quality of the students, the excellence of the faculty and the extraordinary sense of harmony and progress throughout the university.

Since the founding of MIU several other universities bearing Maharishi's name have been established in different parts of the world. In 1976 Maharishi European Research University was founded in Seelisberg, Switzerland, to promote research on the development of higher states of consciousness; in 1982 Maharishi University of Natural law was founded at Mentmore, England, to disseminate the knowledge of the Maharishi Technology of the Unified Field to countries throughout the world. On January 12, 1985, Maharishi Vedic University was inaugurated in Washington, D.C.

MAHARISHI VEDIC UNIVERSITY

Maharishi inaugurated Maharishi Vedic University, the first Vedic university in the world, to bring the support of nature to the individual and the nation through Veda based education. The following are excerpts from Maharishi's inaugural address.

> With great joy and with the fulfillment of all the laudable aspirations of the wise throughout the ages, in the company of great Vedic scholars, experts in different areas of Vedic wisdom, and in the company of great scientists today who are opening a new direction to life, and in the

midst of the Governors of the Age of Enlightenment, it is a very happy time for us all to inaugurate the Vedic University in Washington, D.C., the capital city of the most creative country of the world, where the destiny, efforts and achievements of all mankind are designed day after day.

This is that fortunate city in the world which for the first time in the history of mankind is opening a new gate to education—the field of total knowledge with the infinite organizing power of nature inherent within it. It gives us such great satisfaction at this moment that a new gate to fulfillment for all mankind is being opened in this inauguration of the Vedic University in the capital of this most fortunate and beautiful country.

...This Vedic University that we are now starting here in Washington, D.C., is going to be a milestone of infinite value for all mankind because this university is going to establish a model of education based on the unified field. We are establishing a model of education, a model university where the beginning and end of knowledge—the whole course of knowledge, which means the whole course of the infinite organizing power of nature—will be demonstrated in the life of the students.

...There is already a lot of literature in this Vedic University. We are not creating any new textbooks. All the man-made textbooks are already enough, and they have promoted the objective approach of modern science, which has glimpsed the existence of the unified field. The self-interacting unified field has already been glimpsed, and there are theories from different disciplines of modern science which speak about it. But the Veda in its original script is just the whisper of the unified field to itself, and human awareness, settling down to its own self-referral state, very clearly cognizes its own self-interacting activity.

The four Vedas—Rig Veda, Sama Veda, Yajur Veda, and Atharva Veda—these Vedas are a beautiful, sequentially available script of nature in its own unmanifest state, eternally functioning within itself and on that basis of self-interaction, creating the whole universe and governing it.

...Vedic literature brings out, in detail, the three values of the observer, observed, and observation, where consciousness is open to itself. These three values in the Veda are called rishi, devata, and chhandas. The terminology does not matter; these are the three values, whether we talk in English or Arabic or in the Vedic language itself. These three values are in one. The three values being in one, in this is the pulsating universe. The universe is pulsating—contracting and expanding, contracting and expanding. From where does this phenomenon, which is an eternal phenomenon in nature, come? It comes from the nature of pure consciousness, which is singularity and three together—observer, observed, and observation, all together.

> ...With this technology of the unified field, the Veda, and the knowledge of how and where imbalances begin and of how and from where balance could be restored, it is completely within our reach today to reconstruct society by creating and maintaining a group of 7000 in this city. Through 15 minutes morning and evening regular practice of the Technology of the Unified Field, the students at the Vedic University will gain the knowledge of how nature functions and will spontaneously create coherence in society, national life, and world consciousness.
>
> ...This beautiful university that we are bringing out today is destined to make the country, the continent, and the world self-sufficient, because the U.S.A. is a leading member, a guiding light in the family of nations. We want the guiding light to be so strong that no one in the world is in darkness. This is the lighthouse that we are establishing in this Vedic University today. We have very, very great hope, because this is the country that has proven to be the most generous, and the most advanced country for adopting new things and giving a new lead to the world.

The founding of Maharishi Vedic University has come at a critical time in the history of the world, a time of great imbalance in the functioning of natural law, a time when mankind is under the fear of nuclear destruction. Yet it is also a time of great discovery and revival of ancient wisdom.

As a result of Maharishi's great genius in unfolding the complete knowledge of Vedic Science in the light of the discovery of the unified field of all the laws of nature, we have in our possession the knowledge and technology to restore balance in the world and secure mankind's future. The establishment of Maharishi Vedic University will provide a setting in which to conduct teaching and research at the finest level of nature's functioning, the unified field of all the laws of nature, and will further promote research into the sequential unfoldment of the laws of nature, as described in the ancient Vedic literature.

It will also provide a center for the prevention of disorder and ill health on the level of the individual, society, and world. Through the group practice of the Maharishi Technology of the Unified Field and the introduction of Ayurveda Prevention Centers throughout North America, Maharishi Vedic University will enliven the total potential of natural law in each individual and thus restore balance to world consciousness. As Maharishi said at the conclusion of his inaugural address on January 12, 1985,

> In inaugurating this Vedic University in Washington, we are very full of inner contentment. It is the fulfillment of all these twenty, thirty years of activities throughout the world. We see a new dawn for all mankind in the establishment of this Vedic University.

EDUCATION AND ENLIGHTENMENT—MAHARISHI'S INTEGRATED SYSTEM OF EDUCATION

Maharishi's unified field based integrated system of education is the fulfillment of all the sublime aspirations of educational systems throughout the ages. It connects every part of knowledge to the whole of knowledge, and the whole of knowledge is connected to the Self. At the same time the potential of the conscious mind is fully enlivened. The Maharishi Technology of the Unified Field thus provides the foundation to education by culturing the neurophysiological development of enlightenment.

The creative potential of the Self is developed through every channel. This is the supreme gift of Maharishi's integrated system of education: any knowledge that the student gains from the perspective of the unified field delivers the fruit of all knowledge—freedom from the limitations of boundaries through the full development of his creative potential. In the process of gaining knowledge the student grows in the awareness of the Self as the center of all knowledge.

Because the Self is the lively field of the unified field of all the laws of nature, through this system of education each individual becomes familiar with the total potential of natural law. This enables thought always to be evolutionary and positive so that mistakes are not made. The individual does not pollute the environment by violating natural law; rather, his behavior becomes ideal—he is able to fulfill his own interests without losing sight of the interests of the whole society, and he becomes a citizen of the Age of Enlightenment. The result is a unified field based civilization—as Maharishi has said, "heaven on earth,"—in which each citizen becomes a living example of the highest ideal of life, enlightenment, the full development of neurophysiological functioning.

With Maharishi's integrated system of education, every student becomes a source of the infinite organizing power of nature. Because a university or school is the seat of knowledge in society, all universities and schools collectively should be the source of organizing power and order for the whole society. In the past this has unfortunately not been the case.

With the Maharishi Technology of the Unified Field

every student can now enliven the field of pure knowledge, the unified field of natural law, in his awareness. Thus the whole school or university automatically radiates an influence of organization and order to society. This is the real role of an institution of education. In past years the concept of a university has been *all knowledge on one campus*. Now with Maharishi's integrated system of education, the concept of university has expanded to its level of fulfillment, which is *all knowledge in one human brain*.

Education is the means by which knowledge is perpetuated. The only means to perpetuate the beautiful knowledge Maharishi has given us is through a unified field based educational system in which every individual is given the opportunity to develop his or her full creative potential.

Education is also the means by which nations can become strong. By developing maximum coherence in individual consciousness we can raise the coherence of the collective consciousness of each nation. National law will automatically be supported by the total potential of natural law. Each nation will reflect the highest ideals of life: self-sufficiency, cultural integrity, and invincibility; and peace and harmony will radiate throughout the whole family of nations.

The Maharishi Technology of the Unified Field offers an entirely new approach to life. It provides a completely new principle of action and achievement.

As Maharishi explains,

> Success lies in handling nature, and nature is in one's own nature. The more you are able to handle yourself, the more you are able to handle the world.

> ...This is the wisdom principle. It is the wisdom principle that is going to dominate and create an ideal society and take people out of their attitude of struggling for success. Power is not in action. Power is in knowledge. Because it is knowledge that serves as an engine to action. It is the power of knowledge that propels action. In this we are not undermining the role of action for achievement and success. Success is always born of action. Success and fulfillment come from action. But action is not an independent factor. Action is dependent on knowledge.
>
> The purpose of action is quick achievement. Achievement at the cost of least energy and least time. But, when the actor is not infused with pure knowledge from within himself, he makes mistakes in action, and as a result the whole course of action is full of problems and achievements become difficult. Now at the basis of our global undertaking to create an ideal society, we have changed the philosophy of action. It will mean not using much energy or much time, but the achievement will be greater. This is the practical philosophy that is going to transform the course of action from being shrouded in problems to being the clear sunshine of achievement.

Maharishi's real achievement is to have provided the means not only to create an ideal world, but also to perpetuate an ideal world through unified field based civilization. Ten thousand years from now, when historians look back and marvel at the precarious and dramatic nature of this transition from an age of ignorance to an Age of Enlightenment, how will they regard Maharishi? As a philosopher, a scientist, a Vedic seer, a world leader?

Maharishi's unique qualities certainly encompass all of these, for he is at once an ancient and modern man, the living representative of the oldest and most profound tradition of mankind's search for truth and at the same time the leading scientist into the uncharted regions of the most promising frontier of science: consciousness. His entire life has been devoted to raising the quality of life everywhere, by inspiring the revival of the cultural traditions of each nation and the simultaneous stimulus for nations to adopt a more progressive technology.

I think historians will come to regard Maharishi as the great teacher he has been and continues to be to so many, whether they are leaders or laborers, scholars or school children, scientists or artists—an infinitely wise and compassionate teacher of pure knowledge.

"Maharishi" means great teacher or seer. Maharishi, in this generation, by his living example, sets an ideal for all teachers of knowledge to come. He has with great purity and clarity served as nature's instrument to reawaken in each of us our full potentiality, our own birthright to live in fulfillment and enlightenment.

Like so many others who have shared Maharishi's wisdom, enthusiasm, and joy, I am continually delighted and fulfilled by each new wave of knowledge he brings forth. I sincerely believe that we have only begun to learn a fraction of the wealth of wisdom and knowledge Maharishi has in store for the world, and I cannot help but look forward, with deep gratitude and great excitement, to Maharishi's further teachings of the extraordinary wisdom of Vedic Science and the Maharishi Technology of the Unified Field.

Appendix I

Main Points of

MAHARISHI'S WORLD PLAN

FOR PERFECT HEALTH

1985

INTRODUCTION

Maharishi's World Plan for Perfect Health comes to bring ultimate fulfillment to Maharishi's World Plan formulated in 1972 to solve the age-old problems of mankind in this generation.

It approaches the problems of health from the perspective of prevention, which alone will have a chance to eliminate all pain and suffering from life in all parts of the world.

Maharishi's World Plan for Perfect Health is being presented for implementation by the Ministry of Health and Immortality of the World Government of the Age of Enlightenment to the Ministry of Health of every government.

The basis of this plan is the time-tested knowledge of life, available in the tradition of Ayurveda. The extensive scientific literature of Ayurveda and its preventive and curative preparations and programs have that much needed vitality which promises to eliminate sickness in life and create a world free form sickness and suffering.

As the most ancient system of perfect health, Ayurveda is the source of all traditions of natural medicine in the world. It states that the medicinal plants growing in any country are most effective for maintaining perfect health of the people of that country. It furthermore provides the theoretical framework to classify the national flora of every country and determine their therapeutic action. This makes Ayurveda a universally applicable system of health care.

On the basis of the authentic literature of Ayurveda, it is possible to open training centers of natural medicine in every country.

India has a University of Ayurveda and more than 100 Ayurvedic Colleges throughout the country.

It is the profoundly established system of Ayurveda in India that prompted the World Health Organization to recommend Ayurveda and, on that basis, the indigenous medicine in every country.

Maharishi's World Plan for Perfect Health is designed to bring fulfillment to this new awakening in the world for natural medicine and to bring self-sufficiency in health care to every nation.

THE GOAL

To bring self-sufficiency in
health care to every nation by:

1. establishing Ayurveda Research Centers in every country to verify scientifically the validity of indigenous medicine in the light of the authentic records of Ayurveda;

2. compiling a comprehensive Materia Medica from the medicinal flora of every

country on the basis of comparative studies between the flora of every country and the flora of India, for which a comprehensive Materia Medica is already available in the ancient literature of Ayurveda;

3. exploring scientifically the medicinal value of metals and minerals available in the country and verifying the ancient and traditional procedures to enliven their latent therapeutic and preventive values;

4. establishing Ayurveda Herbal Gardens for the cultivation of medicinal plants;

5. establishing Ayurveda Pharmacies in every country to manufacture natural medicines on the basis of ancient traditional and authentic Ayurvedic recipes under the strictest hygienic conditions, applying the highest international standards of modern scientific quality control at every stage of production;

6. establishing Maharishi Ayurveda Universities and Colleges for the training of Ayurvedic physicians and health professionals in every country;

7. minimizing diseases by popularizing knowledge and programs for the prevention of sickness, preservation of health, and promotion of longevity through Ayurveda Prevention Centers;

8. establishing Ayurveda Clinics and Hospitals in every country for the effective cure of prevailing diseases without creating harmful side effects;

9. implementing Ayurveda Health Education Programs in every country and in all areas of society, including education, health, administration, industry, agriculture, and rehabilitation;

10. caring for the collective health of the nation as a whole by providing every government with the technology to create coherence in collective consciousness and thereby maintain perfect balance in the social and ecological environment, so that all influences of the society on the individual are always health-promoting and evolutionary;

11. establishing, in the first year, one Ayurveda prevention and health care facility in the capital of the country with the help of experts in Ayurveda form India and, in the second and third years when the local Ayurvedic doctors are available, establishing local clinics and prevention centers in all parts of the country according to population density and government programs.

ORGANIZATION

The Ministry of Health and Immortality of the World Government of the Age of Enlightenment and the Ministry of Health of the country will enter into a working relationship and together will accomplish the plan.

The Ministry of Health and Immortality of the World Government of the Age of Enlightenment will provide knowledge and expertise, and the Ministry of Health in the country will provide the infrastructure, preferably through the existing health care policies and programs in the country.

FINANCING

Apart from providing the most effective system of health care for reducing sickness and promoting perfect health and longevity, a striking feature of Maharishi's World Plan for Perfect Health is its extreme cost-effectiveness, which will enable every government to finance the implementation of this plan from the savings it will produce in the national health care budget.

The initial costs of setting up the program, which will amount to only a fraction of the national health budget, will be within the easy reach of any government.

An added advantage of this plan is to reduce excessive expenditure, and trade deficits incurred by the importation of modern medicine, particularly in developing nations. Through this plan it seems possible to reduce the cost of importation of modern medicines by 25% every year for three years, starting from the second year.

International Financing

It will be the joy of every government to receive initial aid from international funding agencies to set up this beautiful program for fulfilling the goals of primary health care for the whole population and thereby raising the country to self-sufficiency in perfect health care.

Funds may also be forthcoming from the multinational manufacturers of modern medicine, as it is obvious that this plan comes as the lfillment of their deep concern and sincere effort to help mankind to come out of suffering.

FULFILLMENT OF THE PLAN

The fulfillment of Maharishi's World Plan for Perfect Health will be realized only when every family in the world has been provided with perfect health care.

As the channels of every government are already set to reach every home and support the life of every family, with the cooperation of the government it will not take more than three to four years to fulfill the plan in any country.

As soon as the training program starts, every day will bring the day of fulfillment nearer.

Solution to All Problems

There is no problem in the world that can be isolated from the problem of health. Every government making the fortunate decision to implement the plan will not only solve the problems of health, but will eliminate the common basis of all problems and bless the nation with the perpetual sunshine of the Age of Enlightenment.

The most fortunate aspect of the plan is that the World Government of the Age of Enlightenment is prepared to undertake the implementation of the plan in all countries simultaneously.

INAUGURATION OF THE PLAN

Maharishi's World Plan for Perfect Health will be inaugurated in each country by the guiding light of the nation, the Head of State, in a nationwide ceremony when the whole nation will rise to celebrate the Head of State as the Founding Father of the Age of Enlightenment, and the World Government of the Age of Enlightenment will honor him as the President of the World Government of the Age of Enlightenment in the country. All members of his governemnt will be honored as the custodians of the Age of Enlightenment in the country.

On 12 January 1986, the plan will be internationally inaugurated by the Governor Generals of the Age of Enlightenment in all the six Continental Capitals of the Age of Enlightenment.

Global Initiative for
Perfect Health

Every Head of State, the guiding light of his nation, is invited to take the lead by inaugurating the plan in his own country and together with all the other Heads of state, on the same date, create a global initiative for perfect health for all mankind.

Throughout history the world family will honor and celebrate 12 January as the "Day of the Founding Fathers of the Age of Enlightenment," and every nation will perpetually enjoy the sunshine of the Age of Enlightenment. Every nation will radiate peace, harmony, and happiness for the whole world family.

Appendix II

Addresses and Contacts

There are several universities and institutes founded by Maharishi which either conduct or support research on Ayurveda.

Maharishi Vedic University
1111 H Street, NW
Washington, D.C. 20005

Maharishi International University
Institute of Ayurveda Studies
Fairfield, Iowa 52556

Maharishi Vedic University
NL-6063 NP Vlodrop, the Netherlands

Maharishi University of Natural Law
Mentmore, Buckinghamshire
England, LU7 0QH

Maharishi European Research University
Seelisberg, Switzerland

Maharishi Ved Vigyan Vidya Peeth
Maharishi Nagar - 201 307, India

Further, the following is a list of the major MAHARISHI AYURVEDA MEDICAL CENTERS:

Maharishi Ayurveda Medical Center
National Office for the United States of America
3201 Glasgow Road
Fairfield, Iowa 52556
515-472-5866

Maharishi Ayurveda Medical Center
Washington, D.C.
2112 F Street NW
Washington, D.C. 20037
202-785-2700

Maharishi Ayurveda Medical Center
National Office for Canada
190 Lees Avenue
Ottawa, Ontario K1S 5L5
613-235-0952

Maharishi Ayurveda Medical Center
Los Angeles
17308 Sunset Boulevard
Pacific Palisades, CA 90272
213-454-5531

Maharishi Ayurveda Medical Center
Boston
106 Main Street
Stoneham, MA 02180
617-662-3230

In addition to these centers the following association of physician interested in Ayurveda has also been formed.

American Association for Ayurvedic Medicine
5000 14th Street NW
Washington, D.C. 20011
202-737-6085

Bibliography

ABOU NADER, T.M. (1986). The effects of an Ayurveda rasayana on aging. Paper presented at Maharishi Vedic University Conference on Ayurveda and Perfect Health, Washington, D.C.

ABOU NADER, T.M., ALEXANDER, C.N., DAVIES, J.L. (1984). The Maharishi Technology of the Unified Field and reduction of armed conflict: A comparative, longitudinal study of Lebanese villages, *in* R.A. Chalmers, G. Clements, H. Schenkluhn, M. Weinless (Eds.): *Scientific Research on the Transcendental Meditation Program: Collected Papers, Vol. 4.* Vlodrop, the Netherlands: MIU Press.

ADEY, W.R., BARVIN, S.M. (1977). Brain interactions with weak electrical and magnetics fields. *Neurosciences Research Program Bulletin* 15: 104-129.

ABRAMS, A.I. (1972). Paired-associate learning and recall: A pilot study of the Transcendental Meditation program, *in* D.W. Orme-Johnson, J.T. Farrow (Eds.): *Scientific Research on the Transcendental Meditation Program: Collected Papers, Vol. 1,* (pp. 377-381). West Germany: MERU Press.

ABRAMS, A.I. (1977). The effects of meditation on elementary school students. *Dissertation Abstracts International* 37(9-A): 5689.

ABRAMS, A.I., SIEGEL, L.M. (1978). The Transcendental Meditation program and rehabilitation at Folsom State Prison: A cross-validation study. *Criminal Justice and Behavior* 5(1): 3-20.

ABRAMS, A.I., SIEGEL, L.M. (1979). Transcendental Meditation and rehabilitation at Folsom Prison: Response to a critique. *Criminal Justice and Behavior* 6(1)): 13-21.

AGARWAL, B.L., KHARBANDA, A. (1979). Effect of Transcendental Meditation on mild and moderate hypertension. *in* R.A. Chalmers, G. Clements, H. Schenkluhn, M. Weinless (Eds.): *Scientific Research on the Transcendental Meditation Program: Collected Papers, Vol. 3.* Vlodrop, the Netherlands: MIU Press.

AKISHIGE, Y. (1968). A historical survey of the psychological studies in Zen. *Kyushu Psychological Studies, V, Bulletin of the Faculty of Literature of the Kyushu University* 11: 1-56.

ALEXANDER, C.N., ABOU NADER, T.M., CAVANAUGH, K.L., DAVIES, J.L., DILLBECK, M.C., KFOURY, R.J., ORME-JOHNSON, D.W. (1984). The effect of the Maharishi Technology of the Unified Field on the war in Lebanon: A time series analysis of the influence of international and national coherence creating assemblies, *in* B. Zeiger (Ed.): *Scientific Research on the Transcendental Meditation Program: Collected Papers, (Vol. 4).* Rheinweiler, West Germany: MERU Press, in press.

ALEXANDER, C.N., CRANSON, R.W., BOYER, R.W., ORME-JOHNSON, D.W. (1986). Transcendental consciousness: A fourth state of consciousness beyond sleep, dreaming and waking, *in* Gackenbach J. (Ed.): *Sourcebook on Sleep and Dreams.* New York, Garland.

ALEXANDER, C.N., DAVIES, J., DIXON, C., OETZEL, R., MUEHLMAN, M. (1986). The Vedic Psychology of human development: A theory of development of higher states of consciousness beyond formal operations, *in* Alexander, C.N., Langer, E., Oetzel, R. (Eds.): *Higher Stages of Human Development: Adult Growth Beyond Formal Operation.* New York, Oxford.

ALEXANDER, C.N., DAVIES, J.L., NEWMAN, R.I.,

CHANDLER, H.M. (1983). The effects of Transcendental Meditation on cognitive and behavioural flexibility, health, and longevity in the elderly: An experimental comparison of the Transcendental Meditation program, mindfulness training, and relaxation, *in* R.A. Chalmers, G. Clements, H. Schenkluhn, M. Weinless (Eds.): *Scientific Research on the Transcendental Meditation Program: Collected Papers, Vol. 4*. Vlodrop, the Netherlands: MIU Press.

ALEXANDER, C.N., GRANT, J., STADTE, C. VON. (1982). The effects of the Transcendental Meditation technique on recidivism: A retrospective archival analysis. Doctoral thesis of first author (summary), Department of Psychology and Social Relations, Havard University, Cambridge, Massachusetts, USA. Also *in* R.A. Chalmers, G. Clements, H. Schenkluhn, M. Weinless (Eds.): *Scientific Research on the Transcendental Meditation Program: Collected Papers, Vol. 4*. Vlodrop, the Netherlands: MIU Press.

ALEXANDER, C.N., MARKS, E.J. (1982). Ego development, personality and behavioural change in inmates practicing the Transcendental Meditation technique or participating in other programs: A summary of cross-sectional and longitudinal results. Doctoral thesis of first author (summary), Department of Psychology and Social Relations, Havard University, Cambridge, Massachusetts, USA. Also *in* R.A. Chalmers, G. Clements, H. Schenkluhn, M. Weinless (Eds.): *Scientific Research on the Transcendental Meditation Program: Collected Papers, Vol. 4*. Vlodrop, the Netherlands: MIU Press.

ALLEN, C.P. (1979). Effects of Transcendental Meditation, electromyographic (EMC) biofeedback relaxation, and conventional relaxation on vasoconstriction, muscle tension, and stuttering: A quantitative comparison, *in* R.A. Chalmers, G. Clements, H. Schenkluhn, M. Weinless (Eds.): *Scientific Research on the Transcendental Meditation Program: Collected Papers, Vol. 4*. Vlodrop, the Netherlands: MIU Press.

ALLISON, J. (1970). Respiratory changes during Transcendental Meditation. *Lancet 1*, 7651: 833-834.

ANAND, B.K., CHHINA, G.S., SINGH, B. (1961). Some aspects of electroencephalographic studies on yogis. *Electroencephalography and Clinical Neurophysiology* 13: 452-456.

ANDERSEN P., ANDERSSON, S.A. (1968). *Physiological basis of the alpha rhythm.* New York: Appleton- Century-Crofts.

APPELLE, S., OSWALD, L.E. (1974). Simple reaction time as a function of alertness and prior mental activity. *Perceptual and Motor Skills* 38: 1263-1268.

ARENANDER. A. (1980). Neurophysiological model of transcending. Paper presented at MIU Neuroscience Conference, Fairfield, Iowa.

ASERINSKY, E., KLEITMAN, N. (1965). Periodic respiratory pattern occurring with eye movements during sleep. *Science* 150: 763-766.

BADAWI, K., WALLACE, R.K., ORME-JOHNSON, D.W., ROUZERE A.M. (1984). Electrophysiologic characteristics of respiratory suspension periods occurring during the practice of the Transcendental Meditation program. *Psychosomatic Medicine* 46(3):267-276.

BAGCHI, B.K., WENGER, M.A. (1957). Electrophysiological correlates of some yogi exercises. *Electroencephalography and Clinical Neurophysiology* Supplement 7: 132-149.

BAKER, R.R. (1978). *The evolutionary ecology of animal migration*. London: Hodder and Stoughton.

BALLOU, D. (1973a). The Transcendental Meditation program at Stillwater Prison, *in* D.W. Orme-Johnson, J.T. Farrow (Eds.): *Scientific Research on the Transcendental*

Meditation Program: Collected Papers, Vol. 1, (pp. 569-576). West Germany: MERU Press.

BALLOU, D. (1973b). TM research: Minnesota State Prison, *in* D.P. Kanellakos & P.C. Ferguson (Eds.): *The Psychobiology of TM.* Los Angeles: MIU Press.

BANQUET, J.P. (1972). EEG and meditation. *Electroencephalography and Clinical Neurophysiology* 33: 454 (Abstract).

BANQUET, J.P. (1973). Spectral analysis of the EEG in meditation. *Electroencephalography and Clinical Neurophysiology* 35: 143-151.

BANQUET, J.P., BOURZEIX, J.C., LESEVRE, N. (1979). Evoked potentials in altered states of consciousness (A.S.C.). Lab. Electrophysiol. Neurophysiol. Appl., Hop. Salpetriere, 75651, Paris. Cedix 13.

BANQUET, J.P., HAYNES, C., HEBERT, J.R., REBER, W. (1977). Analysis of sleep in altered states of consciousness by classical EEG and coherence Spectra. *Electroencephalography and Clinical Neurophysiology* 43(4): 503, E203 (Abstract).

BANQUET, J.P., LESEVRE, N. (1980). Event-related potentials in altered states of consciousness—Motivation and sensory processes of the brain. *Progress in Brain Research* 54: 447-453.

BANQUET, J.P., SAILHAN, M. (1974a). EEG analysis of spontaneous and induced states of consciousness. *Revue d'ELectroencephalographie et de Neurophysiologie Clinique* 4: 445-453.

BANQUET, J.P., SAILHAN, M. (1974b). Quantified EEG spectral analysis of sleep and Transcendental Meditation. Paper presented at Second European Congress on Sleep Research, Rome, *in* D.W. Orme-Johnson, J.T. Farrow

(Eds.): *Scientific Research on the Transcendental Meditation Program: Collected Papers, Vol. 1,* (pp. 182-186). West Germany: MERU Press.

BAUHOFER, U. (1983). Das programm der Transzendentalen Meditation in der Behandlung von Adipositas, *in* R.A. Chalmers, G. Clements, H. Schenkluhn, M. Weinless (Eds.): *Scientific Research on the Transcendental Meditation Program: Collected Papers, Vol. 3.* Vlodrop, the Netherlands: MIU Press.

BEISCHER, D.E. (1971). The null magnetic field as reference for the study of geomagnetic directional effects in animals and man. *Annals of the New York Academy of Science* 188: 324-330.

BENEDICT, F.G., BENEDICT, C.G. (1933). Mental effort in relation to gaseous exchange, heart rate, and mechanics of respiration. Washington, D.C.: Carnegie Institute.

BENNET, E.L., DIAMOND, M.C., KRECH, D., ROSENZUEIG, M.R. (1964). Chemical and anatomical plasticity of the brain. *Science* 146: 610-619.

BENNET, J.E., TRINDER, J. (1977). Hemispheric laterality and cognitive style associated with Transcendental Meditation. *Psychophysiology* 14: 293-296.

BENSON, H., WALLACE, R.K., (1972a). Decreased blood pressure in hypertensive subjects who practiced meditation. *Supplement II to Circulation 45 and 46: 516.*

BENSON, H., WALLACE, R.K. (1972b). Decreased drug abuse with Transcendental Meditation: A study of 1862 subjects, in CJD Zarafonetis (Ed.): *Drug Abuse: Proceedings of the International Conference (pp. 364-376). Philadelphia: Lea and Feibiger.*

BERESFORD, M.S., CLEMENTS, G. (1983). The group

dynamics of consciousness and the U.S. stock market, in R.A. Chalmers, G. Clements, H. Schenkluhn, M. Weinless (Eds.): *Scientific Research on the Transcendental Meditation Program: Collected Papers, Vol. 4.* Vlodrop, the Netherlands: MIU Press.

BERG, W.P. VAN DEN, MULDER, B. (1976). Psychological research on the effects of Transcendental Meditation technique on a number of personality variables, *in* D.W. Orme-Johnson, J.T. Farrow (Eds.): *Scientific Research on the Transcendental Meditation Program: Collected Papers, Vol. 1,* (pp. 428-433). West Germany: MERU Press.

BERGER, H. (1929). Human brain potentials during the onset of sleep. *Archives of Psychiatry* 87: 527-570.

BERKER, E. (1974). Stability of skin resistance responses one week after instruction in the Transcendental Meditation technique, *in* D.W. Orme-Johnson, J.T. Farrow (Eds.): *Scientific Research on the Transcendental Meditation Program: Collected Papers, Vol. 1,* (pp. 243-250). West Germany: MERU Press.

BERNARD, C. (1932). Les Phenomenes de la Vie, *in* W.B. Cannon: *The wisdom of the body.* New York: Norton.

BEVAN, A.J.W. (1980). Endocrine changes in Transcendental Meditation. *Clinical and Experimental Pharmacology and Physiology* 7: 75-76.

BEVAN, A.J.W., YOUNG, P.M., WELLBY, M.L. NENADOVIC, P., DICKINS, J.A. (1976). Endocrine changes in relaxation procedures. *Proceedings of the Endocrine Society of Australia* 19: 59.

BLACKMORE, C.J., COOPER, G.E. (1970). Development of the brain depends on the visual environment. *Nature* 206: 854-856.

BLACKWELL, B., HANENSON, I.B., BLOOMFIELD, S.S., MAGENHEIM, H.G., NIDICH, S.I., GARTSIDE, P. (1975). Effects on Transcendental Meditation on blood pressure: A controlled pilot experiment. *Psychosomatic Medicine* 37(1): 86.

BLASDELL, K. (1977). The Effects of Transcendental Meditation Upon a Complex Perceptual-Motor Task, *in* D.W. Orme-Johnson, J.T. Farrow (Eds.): *Scientific Research on the Transcendental Meditation Program: Collected Papers, Vol. 1,* (pp. 322-325). West Germany: MERU Press.

BLEICK, C.R. (1982). The influence of the Transcendental Meditation program on criminal recidivism, *in* R.A. Chalmers, G. Clements, H. Schenkluhn, M. Weinless (Eds.): *Scientific Research on the Transcendental Meditation Program: Collected Papers, Vol. 3.* Vlodrop, the Netherlands: MIU Press.

BLEICK, C.R., ABRAMS, A.I. (1984). The Transcendental Meditation program and criminal recidivism in California, *in* R.A. Chalmers, G. Clements, H. Schenkluhn, M. Weinless (Eds.): *Scientific Research on the Transcendental Meditation Program: Collected Papers, Vol. 4.* Vlodrop, the Netherlands: MIU Press.

BOHR, N. (1961). *Atomic theory and the description of nature.* Cambridge, England: Cambridge University Press.

BORLAND, C., LANDRITH, G. (1976). Improved quality of city life through the Transcendental Meditation program: Decreased crime rate, *in* D.W. Orme-Johnson, J.T. Farrow (Eds.): *Scientific Research on the Transcendental Meditation Program: Collected Papers, Vol. 1,* (pp. 639-648). West Germany: MERU Press.

BOUMAN, M.A. (1961). History and present status of quantum theory of vision, *in* W.A. Rosenblith (ed.): *Sensory Communication.* Cambridge, Massachusetts: MIT

Press.

BOURLIERE, F. (1970). *The assessment of biological age in man.* Geneva: World Health Organization.

BRAUTIGAM,E. (1972). Effects of the Transcendental Meditation program on drug abuse; A prospective study, *in* D.W. Orme-Johnson, J.T. Farrow (Eds.): *Scientific Research on the Transcendental Meditation Program: Collected Papers, Vol. 1,* (pp. 506-514). West Germany: MERU Press.

BROOKS, J.S., SCARANO,T. (1985). Transcendental Meditation in the treatment of post-Vietnam adjustment. *Journal of Counseling and Development* 64(3): 212-215.

BROSSE, T. (1946). A psychophysiological study. *Main Currents in Modern Thought* 4: 77-84.

BUELL, S., COLEMAN, D. (1979). Dendrite growth in the aged human brain and failure of growth in senile dementia. *Science* 206: 854-856.

BUJATTI, M., RIEDERER, P. (1976). Serotonin, noradrenaline, dopamine metabolites in Transcendental Meditation technique. *Journal of Neural Transmission* 39: 257-267.

BURGMANS, W.H.P.M., BURGT, A.T. VAN DER, LANGENKAMP, F.P.Th., VERSTEGEN, J.H. (1982). Sociological effects of the group dynamics of consciousness: Decrease of crime and traffic accidents in Holland, *in* R.A. Chalmers, G. Clements, H. Schenkluhn, M. Weinless (Eds.): *Scientific Research on the Transcendental Meditation Program: Collected Papers, Vol. 4.* Vlodrop, the Netherlands: MIU Press.

CAJAL Y, R. (1972). *Histologie du systeme nerveux de l'homme et des vertebres.* Madrid: Conseil Superieur des Investigations scientifiques.

CANNON, W.B. (1932). *The wisdom of the body*. New York: Norton.

CAVANAUGH, K.L., ORME-JOHNSON, D.W., GELERLOOS, P. (1984). The effect of the Taste of Utopia Assembly on the World Index of international stock prices, *in* R.A. Chalmers, G. Clements, H. Schenkluhn, M. Weinless (Eds.): *Scientific Research on the Transcendental Meditation Program: Collected Papers, Vol. 4*. Vlodrop, the Netherlands: MIU Press.

CLEMENTS, G., CLEMENTS, D.M. (1980). The Transcendental Meditation and TM-Sidhi programme and the reversal of ageing. Maharishi European Research University, Seelisberg, Switzerland. Rheinweiler, W. Germany: MERU Press.

CLEMENTS, G., MILSTEIN, S.L. (1977). Auditory thresholds in *in* D.W. Orme-Johnson, J.T. Farrow (Eds.): *Scientific Research on the Transcendental Meditation Program: Collected Papers, Vol. 1*, (pp. 719-722). West Germany: MERU Press.

COLLIER, R.W. (1973). The effect of Transcendental Meditation program upon university academic attainment, *in* D.W. Orme-Johnson, J.T. Farrow (Eds.): *Scientific Research on the Transcendental Meditation Program: Collected Papers, Vol. 1*, (pp. 393-395). West Germany: MERU Press.

CONOR, J.P., DIAMOND, M.C., JOHNSON, R.E. (1980). Occipital cortical morphology of the rat: Alteration with age and environment. *Experimental Neurology* 68: 158-170.

COOPER, M.J., AYGEN, M.M. (1978). Effect of Transcendental Meditation on serum cholesterol and blood pressure. *Harefuah, the Journal of the Israel Medical Association* 95(1): 1-2.

COOPER, M.J., AYGEN, M.M. (1979). A relaxation technique in the management of hypercholesterolemia. *Journal of Human Stress* 5(4): 24-27.

COPE, F.W. (1974). Enhancement by high electrical fields of superconduction in organic biological solids at room temperature and a role of nerve conduction? *Physiological Chemistry and Physics* 6:405.

CRICK, F.H. (1979). Thinking about the brain, *in* D. Hubel: *The Brain*. San Francisco: W.H. Freedman and Co.

CUNNINGHAM, M., KOCH, W. (1973). The Transcendental Meditation program and rehabilitation: A pilot project at the federal correctional institution at Lompoc, California, *in* D.W. Orme-Johnson, J.T. Farrow (Eds.): *Scientific Research on the Transcendental Meditation Program: Collected Papers, Vol. 1,* (pp. 562-568). West Germany: MERU Press.

DANIELS, D. (1976). Comparison of the Transcendental Meditation technique to various relaxation procedures, *in* R.A. Chalmers, G. Clements, H. Schenkluhn, M. Weinless (Eds.): *Scientific Research on the Transcendental Meditation Program: Collected Papers, Vol. 2.* Vlodrop, the Netherlands: MIU Press.

DAS, N.N., GASTAUT, H. (1957). Variation de l'activite electrique du cerveau, du coeur et des muscles squelletiques au cours de la meditation et de l'extase yogique. *Electroencephalography and Clinical Neurophysiology,* Suppl. No. 6: 211-219.

DAVIDSON, R.J., GOLEMAN, D.J., SCHWARTZ, G.E., (1976). Attentional and affective concomitants of meditation: A cross-sectional study. *Journal of Abnormal Psychology* 85: 235-238.

DAVIES, J.L., ALEXANDER, C.N. (1983). The Maharishi

Technology of The Unified Field and improved quality of life in the United States: A study of the First World Peace Assembly, Amherst, *in* R.A. Chalmers, G. Clements, H. Schenkluhn, M. Weinless (Eds.): *Scientific Research on the Transcendental Meditation Program: Collected Papers, Vol. 4.* Vlodrop, the Netherlands: MIU Press.

DEMENT, W.C., MASSERMAN (1964). *Science and psychoanalysis: Scientific proceedings of the academy of psychoanalysis.* New York: Grune and Statton.

DHANARAJ, V.H., SINGH, M. (1973). Reduction in metabolic rate during the practice of the Transcendental Meditation technique, *in* D.W. Orme-Johnson, J.T. Farrow (Eds.): *Scientific Research on the Transcendental Meditation Program: Collected Papers, Vol. 1,* (pp. 137-139). West Germany: MERU Press.

DIAMOND, M. (1980). The Aging Brain: Some Enlightening and Optimistic Results. *American Scientist* Jan.-Feb.

DIAMOND, C., INGHAM, C.A., JOHNSON, R.F., BENNET, E.L., ROSENZWEIG, M.R. (1976). Effects of environment on morphology of rat cerebral cortex and hippocampus. *Journal of Neurobiology* 1: 75-86.

DICK, L.D., RAGLAND, R.E. (1973). A study of the Transcendental Meditation program in the service of counseling, *in* D.W. Orme-Johnson, J.T. Farrow (Eds.): *Scientific Research on the Transcendental Meditation Program: Collected Papers, Vol. 1,* (pp. 600-604). West Germany: MERU Press.

DILLBECK, M.C. (1977). The effect of the Transcendental Meditation technique on anxiety level. *Journal of Clinical Psychology* 33(4): 1076-1078.

DILLBECK, M.C. (1982). Meditation and flexibility of visual perception and verbal problem solving. *Memory and Cognition* 10(3): 207-215.

DILLBECK, M.C. (1979). The Transcendental Meditation program and a compound probability model as predictors of crime rate change, *in* R.A. Chalmers, G. Clements, H. Schenkluhn, M. Weinless (Eds.): *Scientific Research on the Transcendental Meditation Program: Collected Papers, Vol. 4.* Vlodrop, the Netherlands: MIU Press.

DILLBECK, M.C., BRONSON, E.C. (1981). Short-term longitudinal effects of the Transcendental Meditation technique on EEG power and coherence. *International Journal of Neuroscience* 14: 147-151.

DILLBECK, M.C., CAVANAUGH, K.L., BERG, W.P. VAN DEN. (1983). The effect of the group dynamics of consciousness on society: Reduced crime in the Union Territory of Delhi, India, *in* R.A. Chalmers, G. Clements, H. Schenkluhn, M. Weinless (Eds.): *Scientific Research on the Transcendental Meditation Program: Collected Papers, Vol. 4.* Vlodrop, the Netherlands: MIU Press.

DILLBECK, M.C., FOSS, A.P.O., ZIMMERMANN, W.J. (1983). Maharishi's Global Ideal Society Campaign: Improved quality of life in Rhode Island through the TM and TM-Sidhi program, *in* R.A. Chalmers, G. Clements, H. Schenkluhn, M. Weinless (Eds.): *Scientific Research on the Transcendental Meditation Program: Collected Papers, Vol. 4.* Vlodrop, the Netherlands: MIU Press.

DILLBECK, M.C., LANDRITH III, G., ORME-JOHNSON, D.W. (1981b). The Transcendental Meditation program and crime rate change in a sample of 48 cities. Findings previously published in *Journal of Crime and Justice* 4: 25-45.

DILLBECK, M.C., LANDRITH III, G.S., POLANZI, C., BAKER, S.R. (1982). The Transcendental Meditation program and crime rate change: A causal analysis, *in* R.A.

Chalmers, G. Clements, H. Schenkluhn, M. Weinless (Eds.): *Scientific Research on the Transcendental Meditation Program: Collected Papers, Vol. 4.* Vlodrop, the Netherlands: MIU Press.

DILLBECK, M.C., LARIMORE, W.E., WALLACE, R.K. (1984a). A time series analysis of the effect of the Maharishi Technology of The Unified Field: Reduction of traffic fatalities in the United States, *in* R.A. Chalmers, G. Clements, H. Schenkluhn, M. Weinless (Eds.): *Scientific Research on the Transcendental Meditation Program: Collected Papers, Vol. 4.* Vlodrop, the Netherlands: MIU Press.

DILLBECK, M.C., MITTLEFEHLDT, V., LUKENBACH, A.P., CHILDRESS, D., ROYER, A., WESTSMITH, L., ORME-JOHNSON, D.W. (1984b). A time series analysis of the relationship between the group practice of the Transcendental Meditation and the TM-Sidhi program and crime rate change in Puerto Rico, *in* R.A. Chalmers, G. Clements, H. Schenkluhn, M. Weinless (Eds.): *Scientific Research on the Transcendental Meditation Program: Collected Papers, Vol. 4.* Vlodrop, the Netherlands: MIU Press.

DILLBECK, M.C., ORME-JOHNSON, D.W., WALLACE, R.K. (1981a). Frontal EEG coherence, H-reflex recovery, concept learning, and the TM-Sidhi program. *International Journal of Neuroscience* 15: 151-157.

DOMASH, L.H. (1975). The Transcendental Meditation technique and quantum physics: Is pure consciousness a macroscopic quantum state
in the brain? *in* D.W. Orme-Johnson, J.T. Farrow (Eds.): *Scientific Research on the Transcendental Meditation Program: Collected Papers, Vol. 1,* (pp. 652-670). West Germany: MERU Press.

DUBEY, G.P., SINGH, R.H. (1970). Human constitution in clinical medicine, *in* K.N. Udupa (Ed.): *Advances in research in indian medicine.* Banaras Hindu University,

India.

DUMERMUTH, G., WALZ, W., SCOLLO-LAVIZZARI, G., KLEINER, B. (1972). Spectral analysis of EEG activity in different sleep stages in normal adults. *European Neurology* 7: 265-296.

ELLIS, G.A. (1979). *Inside Folsom Prison.* ETC Publications, Palm Springs, California.

EPPLEY, K., ABRAMS, A., SHEAR, J. (1985). *The effects of meditation and relaxation techniques: A meta-analysis.* (Prepublication manuscript available from Dr K. Eppley, L-23, Lawrence Livermore National Laboratory, Livermore, California 94550).

EYERMAN, (1981). Transcendental Meditation and mental retardation. *Journal of Clinical Psychiatry* 42(1): 35-36.

FARREL, D.J. (1979). Changes in metabolic rate and heart rate during meditation. *Proc. Nutr. Eoc. Aust.* 4: 136.

FARROW, J.T., HEBERT, J.R. (1982). Breath suspension during the Transcendental Meditation technique. *Psychosomatic Medicine* 44(2): 133-153.

FERGUSON, P.C., GOWAN, J.C. (1976). Psychological findings on Transcendental Meditation. *Journal of Humanistic Psychology* 16(3): 51-60.

FERGUSON, R.E. (1977). The Transcendental Meditation program at Massachusetts Correctional Institution Walpole: An evaluation report, *in* R.A. Chalmers, G. Clements, H. Schenkluhn, M. Weinless (Eds.): *Scientific Research on the Transcendental Meditation Program: Collected Papers, Vol. 2.* Vlodrop, the Netherlands: MIU Press.

FERGUSON, R.E. (1978). A self-report evaluation of the

effects of the Transcendental Meditation program at Massachusetts Correctional Institution Walpole: A follow-up, *in* R.A. Chalmers, G. Clements, H. Schenkluhn, M. Weinless (Eds.): *Scientific Research on the Transcendental Meditation Program: Collected Papers, Vol. 2*. Vlodrop, the Netherlands: MIU Press.

FINCH, C.E. (1977). Neuroendocrine and autonomic aspects of aging, *in* C.E. Finch, L. Hayflick (Eds.): *Handbook of the Biology of Aging*. New York: Van Nostrand and Reinhold.

FINCH, C.E., HAYFLICK, L. (1977). *Handbook of the Biology of Aging*. New York: Van Nostrand and Reinhold.

FREW, D.R. (1974). Transcendental Meditation and productivity. *Academy of Management Journal* 17: 362-368.

FRIEND, K.E. (1975). Effects of the Transcendental Meditation program on work attitudes and behavior, *in* D.W. Orme-Johnson, J.T. Farrow (Eds.): *Scientific Research on the Transcendental Meditation Program: Collected Papers, Vol. 1*, (pp. 630-638). West Germany: MERU Press.

FUSON, J.W. (1976). The effect of Transcendental Meditation on sleeping and dreaming patterns. Doctoral thesis (abbr.), Yale Medical School, New Haven, Connecticut, USA. Also *in* R.A. Chalmers, G. Clements, H. Schenkluhn, M. Weinless (Eds.): *Scientific Research on the Transcendental Meditation Program: Collected Papers, Vol. 2*. Vlodrop, the Netherlands: MIU Press.

GALLOIS, Ph. (1984). Modifications neurophysiologiques et respiratoires lors de la pratique des techniques de relaxation. *L'Encephale* 10: 139-144.

GARNIER, D., CAZABAT, A., THEBAULT, P., GAUGE, Ph. (1984). Etude experimentale de la ventilation pulmonaire pendant la technique de Meditation

Transcendantale: Applications en medecine preventive. *Est Medecine* 4(76): 867.

GAROUTTE, B., AIRD, R.B. (1958). Studies on the cortical pacemaker: Synchrony and a synchrony of bi-laterally recorded alpha and beta activity. *Electroencephalography and Clinical Neurophysiology* 10: 259-268.

GELLHORN, E. (1967). Autonomic Imbalance and the Hypothalamus. University of Minnesota Press.

GEVINS, A.S., SCHAFFER, R.E. (1980). A critical review of electroencephalographic (EEG) correlates of higher cortical functions. *CRC Critical Reviews in Biomedical Engineering* 4: 2.

GLASER, J.L., BRIND, J.L., EISNER, M.J., WALLACE, R.K. (1986). Elevated serum dehydroepiandrasterone sulfate levels in older practitioners of an ayurvedic stress reduction program. Sixteenth Annual Meeting of the Society for Neuroscience, Washington, D.C..

GOLEMAN, D.J., SCHWARTZ, G.E. (1976). Meditation as an intervention in stress reactivity. *Journal of Consulting and Clinical Psychology* 44(3): 456-466.

GORE, S.W., ABRAMS, A., ELLIS, G. (1984). The effect of statewide implementation of the Maharishi Technology of The Unified Field in the Vermont Department of Corrections, *in* R.A. Chalmers, G. Clements, H. Schenkluhn, M. Weinless (Eds.): *Scientific Research on the Transcendental Meditation Program: Collected Papers, Vol. 4.* Vlodrop, the Netherlands: MIU Press.

GRAHAM, J. (1971). The effects of Transcendental Meditation on auditory thresholds, *in* R.A. Chalmers, G. Clements, H. Schenkluhn, M. Weinless (Eds.): *Scientific Research on the Transcendental Meditation Program: Collected Papers, Vol. 2.* Vlodrop, the Netherlands: MIU Press.

HAGELIN, J. S. (1984). Is consciousness the unified quantum field? Maharishi International University, Fairfield, Iowa, USA. Preprint no MIU-THP-012.

HATCHARD, G. (1977). Influence of the Transcendental Meditation programme on crime rate in suburban Cleveland, *in* R.A. Chalmers, G. Clements, H. Schenkluhn, M. Weinless (Eds.): *Scientific Research on the Transcendental Meditation Program: Collected Papers, Vol. 2*. Vlodrop, the Netherlands: MIU Press.

HAYNES, C.T., HEBERT, J.R., REBER, W., ORME-JOHNSON, D.W. (1976). The psychophysiology of advanced participants in the Transcendental Meditation program: Correlations of EEG Coherence, creativity, H-reflex recovery, and experience of transcendental consciousness, *in* D.W. Orme-Johnson, J.T. Farrow (Eds.): *Scientific Research on the Transcendental Meditation Program: Collected Papers, Vol. 1*, (pp. 208-212). West Germany: MERU Press.

HEATON, D.P., ORME-JOHNSON, D. (1975). The Transcendental Meditation program and academic achievement, *in* D.W. Orme-Johnson, J.T. Farrow (Eds.): *Scientific Research on the Transcendental Meditation Program: Collected Papers, Vol. 1*, (pp. 396-399). West Germany: MERU Press.

HEBERT, J.R., LEHMANN, D. (1977). Theta bursts: An EEG pattern in normal subjects practicing the Transcendental Meditation technique. *Electroencephalography and Clinical Neurophysiology* 42: 397-405.

HEIDELBERG, R. (1979). Transzendentale Meditation in der geburtshilflichen Psychoprophylaxe. Doctoral thesis (abbr.), medical Faculty, Free University of Berlin, West Berlin, W. Germany. Also *in* R.A. Chalmers, G. Clements, H. Schenkluhn, M. Weinless (Eds.): *Scientific Research on the Transcendental Meditation Program: Collected*

Papers, Vol. 2. Vlodrop, the Netherlands: MIU Press.

HJELLE, L.A. (1974). Transcendental Meditation and psychological health. *Perceptual and Motor Skills* 39: 623-628.

HOENIG, J. (1968). Medical research on yoga. *Confinia Psychiatrica* 11: 69-89.

HOLT, W.R., CARUSO, J.L., RILEY, J.B. (1978). Transcendental Meditation vs pseudo-meditation on visual choice reaction time. *Perceptual and Motor Skills* 46: 726.

HONSBERGER, R., WILSON, A.F. (1973a). The effects of Transcendental Meditation upon bronchial asthma. *Clinical Research* 22: 278.

HONSBERGER, R., WILSON, A.F. (1973b). Transcendental Meditation in treating asthma. *The Journal of Inhalation Technology* 3(6): 79-80.

HUBEL, D. (1979). *The brain.* San Francisco: W.H. Freedman and Co.

HUBEL, D., WEASLE, T.N. (1979). Brain mechanisms of vision, *in* D. Hubel: *The Brain.* San Francisco, W.H. Freeman and Co.

HUGON, M. (1973). Methodology of the Hoffman reflex in man, *in* J.E. Desmed (Ed.): *New Developments in Electromyography and Clinical Neurophysiology,* vol. 3. Basel: Karger.

JAMES, W. (1910). *The Varieties of Religious Experience: A study in human nature.* New York: Longmans, Green and co.

JEVNING, R., PIRKLE, H.C., WILSON, A.F. (1977). Behavioural alteration of plasma phenylalanine concentration. *Physiology and Behavior* 19: 611-614.

JEVNING, . R., WILSON, A.F. (1978b). Behavioural increases in cerebral blood flow. *The Physiologist* 21(4): 60. (Abstract)

JEVNING, R., WILSON, A.F., DAVIDSON, J.M. (1978c). Adrenocortical activity during meditation. *Hormones and Behavior* 10(1): 54-60.

JEVNING, R., WILSON, A.F., O'HALLORAN, J.P., WALSH, R.N. (1983b). Forearm blood flow and metabolism during stylized and unstylized states of decreased activation. *American Journal of Physiology 245: Regulatory Integrative Comp. Physiol.* 14(1): R110-R116.

JEVNING,R., WILSON, A.F., PIRKLE, H., D'HALLORAN, J.P., WALSH, R.N. (1983a). Metabolic control in a state of decreased activation: Modulation of red cell metabolism. *American Journal of Physiology 245: Cell. Physiol.* 14: C457-C461.

JEVNING, R., WILSON, A.F., SMITH, W.R., MORTON, M.E. (1978a). Redistribution on blood flow in acute hypometabolic behavior. *American Journal of Physiology* 235(4): R89-R92.

JEVNING, R., WILSON, A.F., VANDERLAAN, E.F. (1978d). Plasma prolactin and growth hormone during meditation. *Psychosomatic Medicine* 40(4): 329-333.

JOHN, E.R. (1977). *Functional Neuroscience Vol. 2: Neurometrics, Clinical application of quantitative electrophysiology*. Hillsdale, N.J.: Erlbaum.

JOHN, E.R., KARMEL, B.Z., CORNING, W.C., EASTON, A., BROWN, D., AHN, H., JOHN, M., HARMORY, T.,, PRICHEP, L., TORO, A., GERSON, I., BARLLETT, F., TATCHER, R., KAYE, H., VALDES, P., SCHWARTZ, E.

(1970). Neurometrics: Numerical taxonomy identifies different profiles of brain function within groups of behavioural similar people. *Science* 196: 1393.

JOUVET, M. (1962). Recherches sur les structures nerveuses et les mecanismes responsables des differentes phases du sommeil physiologique. *Archives Italiennes de Biologie* 100: 125-206.

JOUVET, M. (1967). The states of sleep.*Scientific American* 216: 62-72.

KARAMBELKAR, P., BHOLE, M., GHAROTTE, M. (1969). Effect of yogic asanas on uropepsin excretion. *Indian Journal of Medical Research* 57: 944-947.

KASAMATSU, A., HIRAI, T. (1966). An electroencephalographic study of the Zen meditation (Zazen). *Folio Psychiatry Neurology Japonica* 20: 315-336.

KATZ, D. (1974). Decreased drug use and prevention of drug use through the Transcendental Meditation program, *in* D.W. Orme-Johnson, J.T. Farrow (Eds.): *Scientific Research on the Transcendental Meditation Program: Collected Papers, Vol. 1,* (pp. 536-543). West Germany: MERU Press.

KESTERSON, J. (1986). Changes in respiratory control during Transcendental Meditation. Doctoral Dissertation, MIU Library, Fairfield, Iowa.

KOBAL, G., WANDHOFER, A., PLATTIG, K.H. (1975). EEG power spectra and auditory evoked potentials in Transcendental Meditation (TM). *Pflueger's Archiv.* 359: R96.

KOCHABHAKDI, N.J., CHENTANEZ, (1980). Paper presented at International Congress: Research on Higher States of Consciousness. Bangkok Thialand, January 1980.

KORY, R.K., HUFNAGEL, P. (1974). The effect of the science of creative intelligence course on high school students: A preliminary report, *in* D.W. Orme-Johnson, J.T. Farrow (Eds.): *Scientific Research on the Transcendental Meditation Program: Collected Papers, Vol. 1*, (pp. 400-402). West Germany: MERU Press.

KRAS, D.J. (1974). The Transcendental Meditation technique and EEG alpha activity, *in* D.W. Orme-Johnson, J.T. Farrow (Eds.): *Scientific Research on the Transcendental Meditation Program: Collected Papers, Vol. 1*, (pp. 173-181). West Germany: MERU Press.

LANDFORD, A.G. (1984a). The effect of the Maharishi Technology of The Unified Field on stock prices of Washington, D.C. area based corporations, 1980-83: A time series analysis, *in* R.A. Chalmers, G. Clements, H. Schenkluhn, M. Weinless (Eds.): *Scientific Research on the Transcendental Meditation Program: Collected Papers, Vol. 4*. Vlodrop, the Netherlands: MIU Press.

LANDFORD, A.G. (1984b). Reduction in homicide in Washington, D.C. through the Maharishi Technology of The Unified Field, 1980-83: A time series analysis, *in* R.A. Chalmers, G. Clements, H. Schenkluhn, M. Weinless (Eds.): *Scientific Research on the Transcendental Meditation Program: Collected Papers, Vol. 4*. Vlodrop, the Netherlands: MIU Press.

LANDRITH III, G.S., DILLBECK, M.C. (1983). The growth of coherence in society through the Maharishi Effect: Reduced rates of suicides and auto accidents, *in* R.A. Chalmers, G. Clements, H. Schenkluhn, M. Weinless (Eds.): *Scientific Research on the Transcendental Meditation Program: Collected Papers, Vol. 4*. Vlodrop, the Netherlands: MIU Press.

LANG, R., DEHOF, K., MEURER, K., KAUFMANN, W. (1979). Sympathetic activity and Transcendental

Meditation. *Journal of Neural Transmission* 44: 117-135.

LAZAR, Z., FARWELL, L., FARROW, J.T. (1972). The effects of the Transcendental Meditation program on anxiety, drug abuse, cigarette smoking, and alcohol consumption, *in* D.W. Orme-Johnson, J.T. Farrow (Eds.): *Scientific Research on the Transcendental Meditation Program: Collected Papers, Vol. 1,* (pp. 524-535). West Germany: MERU Press.

LEVANDER, V.L., BENSON, H., WHEELER, R.C., WALLACE, R.K. (1972). Increased forearm blood flow during a wakeful hypometabolic state. *Federation Proceedings* 31: 405. (Abstract)

LEVINE, P.H. (1976). The coherence spectral array (COSPAR) and its application to the study of spatial ordering in the EEG. *Proceedings of the San Diego Bio-Medical Symposium* 15: 237-247.

LEVINE, P.H., HEBERT, J.R., HAYNES, C.T., STROBEL, U. (1975). EEG coherence during the Transcendental Meditation technique, *in* D.W. Orme-Johnson, J.T. Farrow (Eds.): *Scientific Research on the Transcendental Meditation Program: Collected Papers, Vol. 1,* (pp. 187-207). West Germany: MERU Press.

LIN, Y.G., CHANDRA (1980). Paper presented at International Congress: Research on Higher States of Consciousness, Bangkok, Thialand, January 1980.

LITTLE, N.A. (1974). The existence of persistent states in the brain. *Mathematical Biosciences* 19: 101-120.

LONDON, F. (1961). *Superfluids, vol. 1. New York: Dover Publishers.*

LOVELL-SMITH, H.D. (1982). Transcendental Meditation—treating the patient as well as the disease. The New Zealand Family Physician 9: 62-65.

LOWN, B., TEMTE, J., REICH, P.L., GAUGHAM, C., REGESTEIN, Q., HAI, H. (1976). Basis for recurring ventricular fibrillation in the absence of coronary heart disease. *England Journal of Medicine* 294(12): 623-629.

MacCULLUM, M.J. (1974). The Transcendental Meditation program and creativity, *in* D.W. Orme-Johnson, J.T. Farrow (Eds.): *Scientific Research on the Transcendental Meditation Program: Collected Papers, Vol. 1,* (pp. 410-414). West Germany: MERU Press.

MAGOUN, H.W. (1963). *The Waking Brain.* Springfield, Ill.: Charles C. Thomas.

MAHARISHI MAHESH YOGI. (1976). *Creating an Ideal Society.* West Germany: MERU Press.

MAHARISHI MAHESH YOGI. (1978). *Enlightenment and Invincibility.* West Germany: MERU Press.

MAHARISHI MAHESH YOGI. (1969). *On the Bhagavad-Gita.* Baltimore: Penguin Press.

MAHARISHI MAHESH YOGI. (1963). *The Science of Being and the Art of Living.* New York: New American Library.

MARCUS, J.B. (1975). The Transcendental Meditation technique: Consciousness expansion as a rehabilitation technique. *Journal of Psychedelic Drugs* 7: 169-179.

MARTINETTI, R.F. (1976). Influence of Transcendental Meditation on perceptual illusion: A pilot study. *Perceptual and Motor* Skills 43: 822.

MARX, J.L. (1985). The immune system "belongs in the body". *Science* 227: 1190-1192.

MASLOW, A.H. (1969). "Theory Z." *Journal of*

Transpersonal Psychology Fall: 31-47.

MASLOW, A.H. (1968). *Toward a Psychology of Being*. New York: Harper and Row.

McCAY, C.M. (1952). Chemical aspects of ageing and the effects of diet upon ageing, *in* A.I. Lansing (Ed.): *Cowdry's problems of aging*. Baltimore: William & William Co.

McCAY, C.M., CROWELL, M.F. (1934). Prolong the lifespan. *Scientific Monthly* 39: 405-414.

McCUAIG, L.W. (1974). Salivary electrolytes, protein, and pH during Transcendental Meditation. *Experientia* 30(9): 988-989.

McDONAGH, J.M., EGEN, T. (1973). The Transcendental Meditation technique and temperature homeostasis, *in* D.W. Orme-Johnson, J.T. Farrow (Eds.): *Scientific Research on the Transcendental Meditation Program: Collected Papers, Vol. 1*, (pp. 261-263). West Germany: MERU Press.

McEVOY, T.M., FRUMPKIN, L.R., HARKINS, S.W. (1980). Effects of meditation on brain stem auditory evoked potentials. *International Journal of Neuroscience* 10(2-3): 165-170.

McINTYRE, M.E., SILVERMAN, F.H., TROTTER, W.D. (1975). Transcendental Meditation and stuttering: A preliminary report. *Perceptual and Motor Skills* 39: 294.

MILES (1964). Oxygen consumption during three yoga-type breathing patterns. *Journal of Applied Physiology* 19: 75-82.

MISKIMAN, D.E. (1972a). The treatment of insomnia by the Transcendental Meditation program, *in* D.W. Orme-Johnson, J.T. Farrow (Eds.): *Scientific Research on the*

Transcendental Meditation Program: Collected Papers, Vol. 1, (pp. 296-298). West Germany: MERU Press.

MISKIMAN, D.E. (1972b). The effect of the Transcendental Meditation program on compensatory paradoxical sleep, *in* D.W. Orme-Johnson, J.T. Farrow (Eds.): *Scientific Research on the Transcendental Meditation Program: Collected Papers, Vol. 1,* (pp. 292-295). West Germany: MERU Press.

MISKIMAN, D.E. (1973). The effect of the Transcendental Meditation program on the organization of thinking and recall (secondary organization), *in* D.W. Orme-Johnson, J.T. Farrow (Eds.): *Scientific Research on the Transcendental Meditation Program: Collected Papers, Vol. 1,* (pp. 385-392). West Germany: MERU Press.

MONAHAN, R.J. (1977). Secondary prevention of drug dependence through the Transcendental Meditation program in metropolitan Philadelphia. *The International Journal of Addiction* 12(6): 729-754.

MORGAN, R.F., FEVENS, S.K. (1972). Reliability of the adult growth examination : A standardized list of individual aging. *Perceptual and Motor Skills* 34: 415-419.

NIDICH, S.I. (1975). A study of the relationship of TM to Kohlberg's stages of moral reasoning. (Doctoral dissertation U. of Cincinnati, 1975). *Dissertation Abstracts International*, 1976, 36(7-A), 4361-4362.

NIDICH, S., SEEMAN, W., SEIBERS, M. (1973). Influence of the transcendental meditation program on state anxiety, *in* D.W. Orme-Johnson, J.T. Farrow (Eds.): *Scientific Research on the Transcendental Meditation Program: Collected Papers, Vol. 1,* (pp. 434-436). West Germany: MERU Press.

NIDICH, S.I., RYNCARZ, R.A., ABRAMS, A.I., ORME-

JOHNSON, D.W., WALLACE, R.K. (1983). Kohlbergian cosmic perspective responses, electroencephalography coherence and Transcendental Meditation and TM-Sidhi program. *Journal of Moral Education* 12(3): 166-173.

O'HALLORAN J.P., JEVNING R., WILSON A.F., SKOWSKY R., WALSH R.N., ALEXANDER C. (1985). Hormonal control in a state of decreased activation: potentiation of arginine vasopressin secretion. *Physiological Behavior* 35: 591-595.

OATES R.M. (1985). *The Maharishi Effect.* Fairfield, Iowa: Christopher, Maclay & Co.

ORME-JOHNSON, D.W. (1973). Autonomic stability and Transcendental Meditation. *Psychosomatic Medicine* 35: 341-349.

ORME-JOHNSON, D.W. (1981). Prison rehabilitation and crime prevention through the Transcendental Meditation and TM-Sidhi program, *in* Leonard J. Hippchen (Ed.): *Holistic approaches to offender rehabilitation,* Chapter 15, Springfield, Illinois: Charles C. Thomas.

ORME-JOHNSON, D.W. (1986). Reduced health insurance utilization through the Transcendental Meditation program. (Prepublication of manuscript available from Dr. David Orme-Johnson, Department of Psychology, MIU Fairfield IA, 52556).

ORME-JOHNSON, D.W., ALEXANDER, C.N., BOUSQUET, J.G. (1985). Impact assessment analysis of the effects of coherence creativity groups on international conflicts, *in* R.A. Chalmers, G. Clements, H. Schenkluhn, M. Weinless (Eds.): *Scientific Research on the Transcendental Meditation Program: Collected Papers, Vol. 4.* Vlodrop, the Netherlands: MIU Press.

ORME-JOHNSON, D.W., ALEXANDER, C.N., DAVIES,

J.L., CHANDLER, H.M., LARIMORE, W.E. (1984a). International peace project in the Middle East: The effect of the Maharishi Technology of the Unified Field, *in* R.A. Chalmers, G. Clements, H. Schenkluhn, M. Weinless (Eds.): *Scientific Research on the Transcendental Meditation Program: Collected Papers, Vol. 4.* Vlodrop, the Netherlands: MIU Press.

ORME-JOHNSON, D.C., CAVANAUGH, K.L., ALEXANDER, C.N., GELDERLOOS, P., DILLBECK, M., LANFORD, A.G., ABOU, NADER, T.M. (1984b). The influence of the Maharishi Technology of The Unified Field on world events and global social indicators: The effects of Taste of Utopia Assembly, *in* R.A. Chalmers, G. Clements, H. Schenkluhn, M. Weinless (Eds.): *Scientific Research on the Transcendental Meditation Program: Collected Papers, Vol. 4.* Vlodrop, the Netherlands: MIU Press.

ORME-JOHNSON, ,D.W., CLEMENTS, G., HAYNES, C.T., BADAOUI, K. (1977). Higher states of consciousness: EEG coherence, creativity, and experiences of the sidhis, *in* D.W. Orme-Johnson, J.T. Farrow (Eds.): *Scientific Research on the Transcendental Meditation Program: Collected Papers, Vol. 1,* (pp. 705-712). West Germany: MERU Press.

ORME-JOHNSON, D.W., DILLBECK, M.C. (1986). Meta-Analysis and theoretical framework of physiological research on the Transcendental Meditation technique. (prepublication manuscript available from Dr David Orme-Johnson, Department of Psychology, MIU ,Fairfield, IA 52556).

ORME-JOHNSON, D.W., DILLBECK, M.C. (1985). Results on the World Peace Assembly on Vedic Science held in Washington, D.C., *in* R.A. Chalmers, G. Clements, H. Schenkluhn, M. Weinless (Eds.): *Scientific Research on the Transcendental Meditation Program: Collected Papers, Vol. 4.* Vlodrop, the Netherlands: MIU Press.

ORME-JOHNSON, D.W., DILLBECK, M.C., BOUSQUET, J.G. (1979). The World Peace Project of 1978: An experimental analysis of achieving world peace through the Maharishi Technology of the Unified Field, *in* R.A. Chalmers, G. Clements, H. Schenkluhn, M. Weinless (Eds.): *Scientific Research on the Transcendental Meditation Program: Collected Papers, Vol. 2.* Vlodrop, the Netherlands: MIU Press.

ORME-JOHNSON, D.W., DILLBECK, M.C., WALLACE, R.K., LANDRITH III, G.S. (1982). Intersubject EEG coherence: Is consciousness a field? *International Journal of Neuroscience* 16: 203-209.

ORME-JOHNSON, D.W., GELDERLOOS, P. (1984). The long-term effects of the Maharishi Technology of the Unified Field on the quality of life in the United States (1960 to 1983), *in* R.A. Chalmers, G. Clements, H. Schenkluhn, M. Weinless (Eds.): *Scientific Research on the Transcendental Meditation Program: Collected Papers, Vol. 4.* Vlodrop, the Netherlands: MIU Press.

ORME-JOHNSON, D.W., GRANIERI, B. (1977b). The effects of the age of enlightenment governor training courses on field independence, creativity, intelligence, and behavioural flexibility, *in* D.W. Orme-Johnson, J.T. Farrow (Eds.): *Scientific Research on the Transcendental Meditation Program: Collected Papers, Vol. 1,* (pp. 713-718). West Germany: MERU Press.

ORME-JOHNSON, D.W., HAYNES, C.T. (1981). EEG phase coherence, pure consciousness, creativity, and TM-Sidhi experiences. *International Journal of Neuroscience* 13: 211-217.

ORME-JOHNSON, D.W., KIEHLBAUCH, J., MOORE, R., BRISTOL, J. (1971). The Transcendental Meditation program and rehabilitation: A pilot project at the Federal Correctional Institution at Lompoc, California, *in* D.W.

Orme-Johnson, J.T. Farrow (Eds.): *Scientific Research on the Transcendental Meditation Program: Collected Papes, Vol. 1*, (pp. 556-561). West Germany: MERU Press.

ORME-JOHNSON, D.W., WALLACE R.K., DILLBECK M.C., ALEXANDER C.N., BALL, O.E. (1981). Improved functional organization of the brain through the Maharishi Technology of the Unified Field as indicated by changes in EEG coherence and its cognitive correlates: A proposed model of higher states of consciousness, *in* R.A. Chalmers, G. Clements, H. Schenkluhn, M. Weinless (Eds.): *Scientific Research on the Transcendental Meditation Program: Collected Papers, Vol. 4*. Vlodrop, the Netherlands: MIU Press.

OSWALD, I. (1962). *Sleeping and waking: Physiology and psychology*. Amsterdam: Elsevier.

PAGANO, R.R., FRUMKIN, L.R. (1977). The effect of Transcendental Meditation on right hemispheric functioning. *Biofeedback and Self-Regulation* 2(4): 407-415.

PALMORE, E. (Ed.) (1974). *Normal Aging II*. Durham: Duke University Press.

PELLETIER, K.R. (1974). Influence of Transcendental Meditation upon autokinetic perception. *Perceptual and Motor Skills* 39: 1031-1034.

PENNER, W.J., ZINGLE, H.W., DYCK, R., TRUCH, S. (1974). Does an in-depth Transcendental Meditation course effect change in the personalities of the participants? *Western Psychologise* 4: 104-111.

PIROT, M. (1973). The effects of the Transcendental Meditation technique upon auditory discrimation, *in* D.W. Orme-Johnson, J.T. Farrow (Eds.): *Scientific Research on the Transcendental Meditation Program: Collected Papers, Vol. 1*, (pp. 331-334). West Germany: MERU Press.

PIVIK, R.T., MERCIER, L. (1979). Motoneuronal excitability during wakefulness and non-REM sleep: H-reflex recovery function in man. *Sleep* 1(4): 357-367.

PRIGOGINE, I., GLANSDORFF, P. (1971). *Thermodynamic theory of structure, stability and fluctuations*. London and New York: Wiley—Interscience.

PRIGOGINE, I., NICOLAS, G., BABLOYANTZ, A. (1972). Thermodynamic of evolution. *Physic Today* Part I, November: 23, Part II, December: 38.

RABINOFF, R.A., DILLBECK, M.C., DEISSLER, R. (1981). Effect of coherent collective conciousness on the weather, *in* R.A. Chalmers, G. Clements, H. Schenkluhn, M. Weinless (Eds.): *Scientific Research on the Transcendental Meditation Program: Collected Papers, Vol. 4*. Vlodrop, the Netherlands: MIU Press.

RAMIREZ, J. (1976). The Transcendental Meditation program as a possible treatment modality for drug offenders: Evaluation of a pilot project at Milan Federal Correctional Institution. Unpublished report, Milan Correction Institution, Milan, Michigan, 1.

RAO, S. (1962). Metabolic cost of head-stand posture. *Journal of Applied Physiology* 17: 117-118.

RAO, S. (1968). Oxygen consumption during yoga-type breathing at altitudes of 520m and 3800m. *Indian Journal of Medical Research* 56: 701-705.

REDDY, M.K. (1976). The role of the Transcendental Meditation programme in the promotion of athletic excellence: Long and short-term effects and their relation to activation theory, *in* R.A. Chalmers, G. Clements, H. Schenkluhn, M. Weinless (Eds.): *Scientific Research on the Transcendental Meditation Program: Collected Papers, Vol. 2*. Vlodrop, the Netherlands: MIU Press.

REDDY, M.K., BAI, J.L., RAO, V.R. (1974). The effects of the Transcendental Meditation program on athletic performance, *in* D.W. Orme-Johnson, J.T. Farrow (Eds.): *Scientific Research on the Transcendental Meditation Program: Collected Papers, Vol. 1,* (pp. 346-358). West Germany: MERU Press.

RIGBY, B. (1977). Enlightenment in world psychiatry. The Transcendental Meditation technique—new light on consciousness. Paper presented at the Sixth World Congress of Psychiatry, Honolulu, Hawaii, USA., 28 August - 3 September 1977, *in* R.A. Chalmers, G. Clements, H. Schenkluhn, M. Weinless (Eds.): *Scientific Research on the Transcendental Meditation Program: Collected Papers, Vol. 2.* Vlodrop, the Netherlands: MIU Press.

RIMOL, A.G. (1974). The Transcendental Meditation technique and its effects on sensory-motor performance, *in* D.W. Orme-Johnson, J.T. Farrow (Eds.): *Scientific Research on the Transcendental Meditation Program: Collected Papers, Vol. 1,* (pp. 326-330). West Germany: MERU Press.

ROSENMAN, R.H., BRAND, R.J., JENKINS, C.D. et al. (1975). Coronary heart disease in the Western collaborative group study. *Journal of the American Medical Association* 233: 872-877.

ROSENMAN, R.H., FRIEDMAN, M. (1975). Neurogenic factors in pathogenesis of coronary heart disease. *Medical Clinics of North America* 59: 269-279.

ROSS, M.H., BRASS, G. (1975). Food preference and length of life. *Science* 190: 165-167.

ROUTT, T.J. (1973). Low normal heart and respiratory rates in individuals practicing the Transcendental Meditation technique, *in* D.W. Orme-Johnson, J.T. Farrow (Eds.): *Scientific Research on the Transcendental Meditation Program: Collected Papers, Vol. 1,* (pp. 256-260). West

Germany: MERU Press.

SACHER, G.A. (1978). Longevity, aging and death: an evolutionary perspective. *The Gerontologist* 18: 2-20.

SHECTER, H. (1975). The Transcendental Meditation program in the classroom: A psychological evaluation, *in* D.W. Orme-Johnson, J.T. Farrow (Eds.): *Scientific Research on the Transcendental Meditation Program: Collected Papers, Vol. 1*, (pp. 403-409). West Germany: MERU Press.

SCHENKLUHN, H., GEISLER, M. (1974). A longitudinal study of the influence of the Transcendental Meditation program on drug abuse, *in* D.W. Orme-Johnson, J.T. Farrow (Eds.): *Scientific Research on the Transcendental Meditation Program: Collected Papers, Vol. 1*, (pp. 544-555). West Germany: MERU Press.

SCHNEIDER, R.H., CAVANAUGH, K., ROTHENBERG, S., AVERBACH, R., WALLACE, R.K. (1985b). Improvements in health with the Maharishi Ayurveda Prevention Program. Paper presented at the Eight World Congress of the International College of Psychosomatic Medicine, Chicago.

SCHNEIDER, R.H., ORME-JOHNSON, D., KESTERSON, J., WALLACE, R.K. (1985a). Physiological and psychological correlates of Ayurvedic psychosomatic types. Paper presented at Eight World Congress of the International College of Psychosomatic Medicine, Chicago.

SCHRODINGER. E. (1945). *What is life?* London: Cambridge University Press.

SCHWARTZ, E. (1979). The effects of the Transcendental Meditation program on strength of the nervous system, perceptual reactance, reaction time, and auditory threshold, *in* R.A. Chalmers, G. Clements, H. Schenkluhn, M. Weinless (Eds.): *Scientific Research on the Transcendental*

Meditation Program: Collected Papers, Vol. 3. Vlodrop, the Netherlands: MIU Press.

SEEMAN, W., NIDICH, S., BANTA, T. (1972). Influence of Transcendental Meditation on a measure of self-actualization. *Journal of Counseling Psychology* 19: 184-187.

SHAFII, M., LAVELY, R.A., JAFFE, R.D. (1975). Meditation and the prevention of alcohol abuse. *American Journal of Psychiatry* 131: 942-945.

SHAFII, M., LAVELY, R., JAFFE, R. (1976). Decrease in cigarette smoking following Transcendental Meditation, *in* R.A. Chalmers, G. Clements, H. Schenkluhn, M. Weinless (Eds.): *Scientific Research on the Transcendental Meditation Program: Collected Papers, Vol. 3.* Vlodrop, the Netherlands: MIU Press. Findings previously published in *Meru Journal* 24: 29 (Abstract).

SHAPIRO, J. (1974). The relationship of the Transcendental Meditation program to self-actualization and negative personality characteristics, *in* D.W. Orme-Johnson, J.T. Farrow (Eds.): *Scientific Research on the Transcendental Meditation Program: Collected Papers, Vol. 1,* (pp. 462-467). West Germany: MERU Press.

SHARMA, P.V. (1981). *Caraka-Samhita, vols. 1 and 2.* Varanass, India: Chankhambha Orientalia.

SHAW, R., KOLB, D. (1971). Reaction time following the Transcendental Meditation technique, *in* D.W. Orme-Johnson, J.T. Farrow (Eds.): *Scientific Research on the Transcendental Meditation Program: Collected Papers, Vol. 1,* (pp. 309-311). West Germany: MERU Press.

SIEGEL, L.M. (1981). The Transcendental Meditation program and the treatment of drug abuse, *in* J.H. Lowinson & P. Ruiz (Eds.): *Substance abuse: Clinical problems and perspective.* Baltimore: Williams & Wilkin.

SIMON, D.B., OPARIL, S., DIMBALL, C.P. (1974). The Transcendental Meditation program and essential hypertension, *in* D.W. Orme-Johnson, J.T. Farrow (Eds.): *Scientific Research on the Transcendental Meditation Program: Collected Papers, Vol. 1,* (pp. 268-269). West Germany: MERU Press.

SINGH, B. (1984). Patients and practitioners of Transcendental Meditation. *Psychosomatic Medicine* 4: 347-362.

SINGH, R.H., SINGH, M.B., UDUPA, K.N. (1984). A study of Tridosha as Neurohumors. *Journal of Research in Ayurveda and Sidaha* 1: 1-20.

SMITH, T.R. (1974). The TM technique and skin resistance response to loud tones, *in* D.W. Orme-Johnson, J.T. Farrow (Eds.): *Scientific Research on the Transcendental Meditation Program: Collected Papers, Vol. 1,* (pp. 243-250). West Germany: MERU Press.

STREHLER, B.L. (1977). *Time, cells, and aging.* New York: Academic Press.

STRYKER, T., WALLACE, R.K., ROTHENBERG, S., AVERBACH, R. (1985). Reduction in biological age through an Ayurvedic treatment program. Maharishi International University, Fairfield, Iowa, USA. Paper presented at the Eighth World Congress of the International College of Psychosomatic Medicine, Chicago, USA.

STUART, C.I.J.M., TAKAHASHI, Y., UNEZAWA, H. (1979). Mixed system brain dynamics: Neural memory as a macroscopic ordered state. *Foundations of Physics* 9: 310-318.

SUBRAHMANYAM, S., PORKODI, K. (1980). Neurohumoral correlates of Transcendental Meditation. *Journal of Biomedicine* 1: 73-88.

SUGI, Y., AKUTSU, K. (1968). Studies on respiration and energy-metabolism during sitting in Zazen. *Research Journal of Physical Education* 12: 190-206.

SULLIVAN, C.E. (1980). Breathing in sleep, *in* J. Orem, C.D. Barnes (Eds.): *Physiology in Sleep*. New York: Academic Press.

SUURKULA, J. (1977). The Transcendental Meditation technique and the prevention of psychiatric illness, *in* R.A. Chalmers, G. Clements, H. Schenkluhn, M. Weinless (Eds.): *Scientific Research on the Transcendental Meditation Program: Collected Papers, Vol. 2*. Vlodrop, the Netherlands: MIU Press.

SYNDER, F., HOBSON, J.A., MORRISON, D.F., GOLDFRANK, F. (1964). Changes in respiration, heart rate, and systolic blood pressure in human sleep. *Journal of Applied Physiology* 19: 417-422.

The Review Panel on Coronary—prone Behavior and Coronary Heart Disease (1981). Coronary—prone behavior and coronary heart disease: A critical review. *Circulation* 63(6): 1199-1215.

TENNYSON, A. (1910). In W. James: *The variety of Religious Experiences: A study in human nature*. New York: Longmans, Green and co.

THOREAU, H.D. (1942). *Walden*. New York: Walter J. Black Inc.

THROLL, D.A. (1982). Transcendental Meditation and Progressive relaxation: Their physiological effects. *Journal of Clinical Psychology* 38(3): 522.

TIMIRAS, S. (1978). Biological perspectives on aging. *American Scientist* 66: 605-613.

TJOA, A. (1975). Meditation, neuroticism and intelligence: A follow-up. *Behavior Journal of Psychology* 3: 167-182.

TOOMEY, M, PENNINGTON, B., CHALMERS, R., CLEMENTS, G. (1983a). The practice of the Transcendental Meditation and TM-Sidhi program reverses the physiological ageing process, *in* R.A. Chalmers, G. Clements, H. Schenkluhn, M. Weinless (Eds.): *Scientific Research on the Transcendental Meditation Program: Collected Papers, Vol. 3.* Vlodrop, the Netherlands: MIU Press.

TOOMEY, M., CHALMERS, R., CLEMENTS, G. (1983b). The Transcendental Meditation and TM-Sidhi programme reverses the physiological ageing process: A longitudinal study, *in* R.A. Chalmers, G. Clements, H. Schenkluhn, M. Weinless (Eds.): *Scientific Research on the Transcendental Meditation Program: Collected Papers, Vol. 3.* Vlodrop, the Netherlands: MIU Press.

TOURENNE, C. (1985). A model of the electromagnetic field of th brain at EEG and microwave frequencies. *Journal of Theor. Biology* 116: 495-507.

TRAVIS, F. (1979). Creative thinking and the Transcendental Meditation technique. *The Journal of Creative Behavior* 13(3): 169-180.

UDUPA, K.N., SINGH, R.H., DUBEG, G.P., RAI, V., SINGH, M.B. (1975). Biochemical basis of psychosomatic constitution. *Indian Journal of Medical Research* 63: 923.

VAKIL, R. (1950). Remarkable feat of endurance of a yogi priest. *Lancet* 2: 871.

VAILLANT, G.E. (1979). Natural history of male psychologic health. *New England Journal of Medicine* 301: 1249-1254.

VAN BOXTEL, A. (1976). The relation between monosynergistic spinal reflex amplitudes and some EEG alpha activity parameters. *Electroencephalography and Clinical Neurophysiology* 40: 297-305.

WALKER, E.H. (1970). The nature of consciousness. *Mathematical Biosciences* 7: 131.

WALLACE, R.K. (1970b). Physiological effects of Transcendental Meditation. *Science* 167: 1751-1754.

WALLACE, R.K. (1970a). Physiolocigal effects of Transcendental Meditation technique: A proposed fourth major state of consciousness. Ph.D. thesis, Physiology Department, University of California, Los Angeles.

WALLACE, R.K., BENSON, H. (1972). The physiology of meditation. *Scientific American* 226(2): 84-90.

WALLACE, R.K., BENSON, H., WILSON, A.F. (1971). A wakeful hypometabolic physiologic state. *American Journal of Physiology* 221(3): 795-799.

WALLACE, R.K., BENSON, H., WILSON, A.F., GARRETT, M.D. (1971). Decreased blood lactate during Transcendental Meditation. *Federation Proceedings* 30: 376 Abstract.

WALLACE, R.K., DILLBECK, M.C., JACOBE, E., HARRINGTON, B. (1982b). The effects of the Transcendental Meditation and TM-Sidhi program on the aging process. *International Journal of Neuroscience* 16: 53-58.

WALLACE, R.K., MILLS, P., ORME-JOHNSON, D.W., DILLBECK, M.C. (1983b). Modification of the paired-H reflex through the Transcendental Meditation and TM-Sidhi program. *Experimental Neurology* 79: 77-86.

WALLACE, R.K., MILLS, P.J., ORME-JOHNSON, D.W.,

DILLBECK, M.C., JACOBE, E. (1982a). The paired H reflex and its correlation with EEG coherence and academic performance in normal subjects practicing meditation. *Society for Neuroscience Abstracts* 8: 537.

WALLACE, R.K., SILVER, J., MILLS, P., DILLBECK, M.C., WAGONER, D.E. (1983a). Systolic blood pressure and long-term practice of the Transcendental Meditation and TM-Sidhi program: Effects of TM on systolic blood pressure. *Psychosomatic Medicine* 45(1): 41-46.

WALTER, D.O., RHODES, J.M., BROWN, D., ADEY, W.R. (1966). Comprehensive spectral analysis of human EEG generators in posterior cerebral regions. *Electroencephalography and Clinical Neurophysiology* 20: 224-237.

WALTER, D.O., KADO, R.T., RHODES, J.M., ADEY, W.R. (1967a). Electroencephalographic baseline in astronaut candidates estimated by computation and pattern recognition techniques. *Aerospace Medicine* 38: 271.

WALTER, D.O., RHODES, J.M., ADEY, W.R. (1967b). Discriminatory among states of consciousness by EEG measurements; A study of four subjects. *Electroencephalography and Clinical Neurophysiology* 22: 22.

WALTON, K.G., LEROM, M., SALERNO, J., WALLACE, R.K. (1981). Practice of the Transcendental Meditation (TM) and TM-Sidhi program may affect the circadian rhythm of urinary 5-hydroxyindole excretion. *Society for Neuroscience Abstracts* 7: 48.

WALTON, K.G., MCCORKLE, T., HAUSER, T., MACLEAN, L., WALLACE, R.K., IENI, J., MEYERSON, L.R., (1986). "Substance M" a serotonin modulator candidate from human urine? *in* Y.H. Ehrlich, R.H. Lenox, E. Korecki, W. Berry: *Molecular Mechanisms of Neuronal Responsiveness: A volume of advances in experimental medicine and biology*. Plenum Press, in press.

WANDHOFER, A., KOBAL, G., PLATTIG, K.H. (1976). Shortening of latencies of human auditory evoked brain potentials during the Transcendental Meditation technique. *Zeitschrift EEG-EMG* 7: 99-103.

WANDHOFER, A., PLATTIG, K.H. (1973). Stimulus-linked DC-shift and auditory evoked potentials in Transcendental Meditation (TM). *Pfluegers Archiv.* 343, R79. (Abstract)

WARD, M.C., CHESNEY M.A., SWAN, G.E., BLACK, G.W., PARKER, F.D., ROSENMAN, R.H. (1986). Cardiovascular responses in type A and B men to a series of stressors. *Journal of Behavioural Medicine* 9(1): 43-49.

WARSHAL, D. (1980). Effects of the Transcendental Meditation technique on normal and Jendrassik reflex time. *Perceptual and Motor Skills* 50: 1103-1106.

WEINBERG, S. (1967). A model of leptons. *Physic Review Letters* 19(21): 1264.

WELDON, J.T., ARON, A. (1974). The Transcendental Meditation program and normalization of weight, *in* D.W. Orme-Johnson, J.T. Farrow (Eds.): *Scientific Research on the Transcendental Meditation Program: Collected Papers, Vol. 1,* (pp. 301-306). West Germany: MERU Press.

WENGER, M.A., BAGCHI, B.K. (1961). Studies of autonomic functioning in practitioners of Yoga in India. *Behavioural Science* 6: 312-323.

WENGER, M.A., BAGCHI, B.K., ANAND, B.K. (1961). Experiments in India on "voluntary" control of the heart and pulse. *Circulation* 24: 1319-1325.

WERNER, O., WALLACE, R.K., CHARLES, B., JANSSEN, G., STRYKER, T., CHALMERS, R. (1986). Long-term endocrinologic changes in subjects practicing the Transcendental Meditation and TM-Sidhi Program. *Psychosomatic Medicine* 48(1-2): 59-66.

WESCOTT, M. (1973). Hemispheric symmetry of the EEG during the Transcendental Meditation technique, *in* D.W. Orme-Johnson, J.T. Farrow (Eds.): *Scientific Research on the Transcendental Meditation Program: Collected Papers, Vol. 1,* (pp. 160-164). West Germany: MERU Press.

WIGNER, E.P. (1962). Remarks on the mind/body question, *in* I.J. Good (Ed.): *The Scientist Speculates.* New York: Basic Books Inc.

WIGNER, E.P. (1967). *Symmetries and reflections.* Bloomington: Indiana University Press.

WILCOX, G. (1973). Autonomic functioning in subjects practicing the TM technique, *in* D.W. Orme-Johnson, J.T. Farrow (Eds.): *Scientific Research on the Transcendental Meditation Program: Collected Papers, Vol. 1,* (pp. 239-242). West Germany: MERU Press.

WILLIAMS, P.W., WEST,M. (1975). EEG responses to photic stimulation in persons experienced at meditation. *Electroencephalography and Clinical Neurophysiology* 39: 519-522.

WILSON. A.F. (1984). Changes in forearm metabolism during hypometabolic states. Presented at MIU Neuroscience Conference, Fairfield, Iowa.

WILSON, A.F., HONSBERGER, R., CHIU, J.T., NOVEY, H.S. (1975). Transcendental Meditation and asthma. *Respiration* 32: 74-80.

WOLKOVE, N., KREISMAN, H., DARRAGH, D., COHEN, C., FRANK, H. (1984). Effect of Transcendental Meditation on breathing and respiratory control. *Journal of Applied Physiology: Respiratory, Environmental and Exercise*

Physiology 56(3): 607-612.

WOOD, M.F. (1981). The effectiveness of Transcendental Meditation as a means of improving the echolalic behavior of an autistic student, *in* R.A. Chalmers, G. Clements, H. Schenkluhn, M. Weinless (Eds.): *Scientific Research on the Transcendental Meditation Program: Collected Papers, Vol. 3.* Vlodrop, the Netherlands: MIU Press.

WORDSWORTH, W. (1974). *The prose works of Williams Wordsworth (vol.III).* Oxford: Clarendon Press.

YEE, A.C., DISSANAYAKE, A.S. (1980). Glucose tolerance and the Transcendental Meditation programme (a pilot study), *in* R.A. Chalmers, G. Clements, H. Schenkluhn, M. Weinless (Eds.): *Scientific Research on the Transcendental Meditation Program: Collected Papers, Vol. 3.* Vlodrop, the Netherlands: MIU Press.

ZAMARRA, J.W., BESSEGHINI, I., WITTENBERG, S. (1975). The effects of the Transcendental Meditation program on the exercise performance of patients with angina pectoris, *in* D.W. Orme-Johnson, J.T. Farrow (Eds.): *Scientific Research on the Transcendental Meditation Program: Collected Papers, Vol. 1,* (pp. 270-278). West Germany: MERU Press.

ZUROFF, D.C., SCHWARTZ, J.C. (1978). Effects of TM and muscle relaxation on trait anxiety, maladjustment, locus of control, and drug abuse. *Journal of Consulting and Clinical Psychology* 46(2): 264-271.

ABOUT THE AUTHOR

Dr. Robert Keith Wallace is President of Maharishi Vedic University, Washington D.C. and Co-Director of the doctoral programs in Neuroscience and Physiology at Maharishi International University, Fairfield, Iowa. He is also Director General of Maharishi Ayurveda Prevention Council and Chairman of the Board of Maharishi Ayurvedic Corporation of America. In 1967 he received his B.S. in physics and in 1970 his Ph.D. in physiology both from the University of California at Los Angeles. He did postdoctoral research at Harvard Medical School and from 1971 to 1977 he served as founding president of Maharishi International University (MIU). Dr. Wallace is also professor of Physiology and Chairman of the Department of Physiological and Biological Sciences at MIU. He has published many articles in such leading scientific journals as *Science, The American Journal of Physiology and Scientific American* and continues to conduct research on the physiology of higher states of consciousness and the application of Ayurveda to health and longevity.